Armin Iske

Multiresolution Methods in Scattered Data Modelling

Armin Iske
Lehrstuhl Numerische Mathematik
und Wissenschaftliches Rechnen
Technische Universität München
Boltzmannstr. 3
85747 Garching, Germany
e-mail: iske@ma.tum.de

Cataloging-in-Publication Data applied for

A catalog record for this book is available from the Library of Congress.

Bibliographic information published by Die Deutsche Bibliothek
Die Deutsche Bibliothek lists this publication in the Deutsche Nationalbibliografie;
detailed bibliographic data is available in the Internet at <http://dnb.ddb.de>.

Mathematics Subject Classification (2000): primary: 65Dxx, 65D15, 65D05, 65D07, 65D17, 65Mxx, 65M25, 65M12, 65M50, 65Yxx, 65Y20; secondary: 76Mxx 76M25, 76M28, 76Txx, 76T99

ISSN 1439-7358
ISBN 3-540-20479-2 Springer-Verlag Berlin Heidelberg New York

Springer-Verlag is a part of Springer Science+Business Media
springeronline.com

Printed in Germany

Cover Design: Friedhelm Steinen-Broo, Estudio Calamar, Spain
Cover production: *design & production*
Typeset by the author using a Springer TEX macro package

Printed on acid-free paper 46/3142/LK - 5 4 3 2 1 0

Für Martina, Meira und Marian.

Preface

This application-oriented work concerns the design of efficient, robust and reliable algorithms for the numerical simulation of multiscale phenomena. To this end, various modern techniques from scattered data modelling, such as splines over triangulations and radial basis functions, are combined with customized adaptive strategies, which are developed individually in this work.

The resulting multiresolution methods include thinning algorithms, multilevel approximation schemes, and meshfree discretizations for transport equations. The utility of the proposed computational methods is supported by their wide range of applications, such as image compression, hierarchical surface visualization, and multiscale flow simulation.

Special emphasis is placed on comparisons between the various numerical algorithms developed in this work and comparable state-of-the-art methods. To this end, extensive numerical examples, mainly arising from real-world applications, are provided.

This research monograph is arranged in six chapters:

1. *Introduction*;
2. *Algorithms and Data Structures*;
3. *Radial Basis Functions*;
4. *Thinning Algorithms*;
5. *Multilevel Approximation Schemes*;
6. *Meshfree Methods for Transport Equations.*

Chapter 1 provides a preliminary discussion on basic concepts, tools and principles of multiresolution methods, scattered data modelling, multilevel methods and adaptive irregular sampling. Relevant algorithms and data structures, such as triangulation methods, heaps, and quadtrees, are then introduced in Chapter 2.

Chapter 3 is devoted to radial basis functions, which are well-established and powerful tools for multivariate interpolation and approximation from scattered data. Radial basis functions are therefore important basic tools in scattered data modelling. In fact, various computational methods, which are developed in this work, essentially rely on radial basis function methods.

In Chapter 3, basic features of scattered data interpolation by radial basis functions are first explained, before more advanced topics, such as optimal recovery, the uncertainty relation, and optimal point sampling, are addressed. Moreover, very recent results concerning the approximation order and the numerical stability of polyharmonic spline interpolation are reviewed, before least squares approximation by radial basis functions is discussed.

The extensive discussion in Chapter 4 is devoted to recent developments concerning thinning algorithms with emphasis on their application to terrain modelling and image compression. In their application to image compression, adaptive thinning algorithms are combined with least squares approximation and a customized coding scheme for scattered data. This yields a novel concept for image compression. As confirmed by various numerical examples, this image compression method often gives better or comparable compression rates to the well-established wavelet-based compression method SPIHT.

In Chapter 5, various alternative multilevel approximation methods for bivariate scattered data are discussed. Starting point of this discussion is our multilevel interpolation scheme of [72], which was the first to combine thinning algorithms with scattered data interpolation by compactly supported radial basis functions. Recent improvements of the multilevel method of [72] are discussed in Chapter 5, where a new adaptive domain decomposition scheme for multilevel approximation is proposed. The performance of the various multilevel approximation schemes is compared by using one real-world model problem concerning hierarchical surface visualization from scattered terrain data.

Chapter 6 is concerned with meshfree methods for transport equations. Meshfree methods are recent and modern discretization schemes for partial differential equations. In contrast to traditional methods, such as finite differences (FD), finite volumes (FV), and finite element methods (FEM), meshfree methods do not require sophisticated data structures and algorithms for mesh generation, which is often the most time-consuming part in mesh-based simulations. Moreover, meshfree methods are very flexible and particularly useful for modelling multiscale phenomena.

In Chapter 6, a new adaptive meshfree method of characteristics, called **AMMoC**, is developed for multiscale flow simulation. The advection scheme **AMMoC** combines an adaptive version of the well-known semi-Lagrangian method with local meshfree interpolation by radial basis functions. The good performance of the particle-based method **AMMoC** for both linear and nonlinear transport problems is shown. This is done by using one model problem concerning tracer transportation in the artic stratosphere, and the popular five-spot problem from hydrocarbon reservoir simulation. The latter is concerning two-phase flow in porous media. In this model problem, our meshfree advection method **AMMoC** is compared with two leading commercial reservoir simulators, ECLIPSE and FrontSim of Schlumberger.

Acknowledgement

I wish to take this opportunity in order to thank those of my friends and colleagues who gave me their precious and long-lasting support during my academic career, and the non-academic detour.

First and foremost, it is my great pleasure to express my gratitude to Nira Dyn (Tel-Aviv), Hans-Georg Feichtinger (Vienna), and Thomas Sonar (Braunschweig) for their invaluable scientific and practical advice as well as for their generous encouragement, which was always of great help in many difficult situations, especially when I returned to academia. On this occasion, I thank Folkmar Bornemann for he offered me, just in time, an academic position, equipped with sufficient academic freedom.

I used this freedom in order to acquire two different European research and training networks, MINGLE (Multiresolution in Geometric Modelling), and NetAGES (Network for Automated Geometry Extraction from Seismic), both of which are currently funded by the European Commission, contract no. HPRN-CT-1999-00117 (MINGLE) and IST-1999-29034 (NetAGES).

Due to my involvement in MINGLE and NetAGES, several exciting joint research initiatives with academic and industrial partners were created, which helped to foster the outcome of this work. Especially the intense and very pleasant collaboration with my MINGLE postdoc student, Laurent Demaret, on image compression, and with my NetAGES PhD student, Martin Käser, on meshfree flow simulation, is very fruitful. I wish to thank both of them explicitly for their professional attitude and everlasting drive during our entire collaboration.

But I also wish to thank my colleague Jörn Behrens for he initiated our joint work on meshfree semi-Lagrangian advection schemes. Moreover, the kind invitation from Jeremy Levesley on a research visit at the University of Leicester (UK) in February/March 2001 is gratefully appreciated. Parts of our very exciting collaboration on multilevel approximation schemes were initiated during that visit, which was supported by the UK Research Council through the EPSRC grant GR/RO7769.

Useful assistance with the implementation of the various algorithms and with the preparation of the required numerical experiments was provided by the computer science students Stefan Pöhn, Konstantinos Panagiotou, Bertolt Meier, Eugene Rudoy, Georgi Todorov, and Rumen Traykov (all at Munich University of Technology).

Last but not least, the friendly and effective collaboration with Springer-Verlag, Heidelberg, through Martin Peters and Thanh-Ha Le Thi is kindly appreciated.

I dedicate this book to my wife Martina and our children, Meira and Marian, not only in gratitude for their great understanding and patience during the uncounted hours that were spent on this book.

Munich, November 2003 *Armin Iske*

Acknowledgments

Table of Contents

1 Introduction

This introductory chapter explains basic principles and concepts of *multiresolution methods in scattered data modelling*, being the theme of this work. This modern interdisciplinary research field is currently subject to rapid development, driven by its wide range of applications in various disciplines of computational science and engineering, including geometric modelling and visualization, image and signal processing, and meshfree simulations for modelling multiscale phenomena.

The outline of this chapter is as follows. Important features and key techniques of *scattered data modelling* are explained in Section 1.1, followed by a discussion on *multiresolution methods* in Section 1.2. Basic concepts of multiresolution methods in scattered data modelling include both *multilevel methods*, subject of Section 1.3, and *adaptive irregular sampling*, to be discussed in Section 1.4. In this introduction, relevant tools and techniques are briefly addressed, pointers to the literature of related methods are provided, and a short outline to the following material in this work is given.

1.1 Scattered Data Modelling

Scattered data modelling is concerned with the approximation of mathematical objects by using samples taken at an unorganized set of discrete points, a scattered *point cloud*. The mathematical object may for instance be the boundary of a solid body, the graph of a scalar field, or the solution of a partial differential equation.

In all particular cases considered in this work, the mathematical object is a real-valued function $f : \mathbb{R}^d \to \mathbb{R}$ in d real variables, where $d \geq 1$ is the dimension of the Euclidean space $\mathbb{R}^d$. Moreover, the point cloud is a finite set $X = \{x_1, \ldots, x_N\} \subset \mathbb{R}^d$ of pairwise distinct points. In any case, the modelling relies essentially on information (sample values) carried by points, and no assumptions concerning the spatial distribution of the points are made.

One traditional branch of scattered data modelling is *scattered data fitting*, where f is to be recovered from a given data vector $f\big|_X = (f(x_1), \ldots, f(x_N))$ of function values. This requires selecting a specific approximation scheme, which computes a *suitable* function $s : \mathbb{R}^d \to \mathbb{R}$ from the given input data $f\big|_X$, such that s approximates f *reasonably* well. One obvious way for doing so is to consider using *scattered data interpolation*, in which case the approximant s is required to satisfy the interpolation conditions $s\big|_X = f\big|_X$, i.e.,

$$f(x_j) = s(x_j), \qquad \text{for } 1 \le j \le N.$$

The function s is in this case referred to as an *interpolant* to f at X.

The subject of scattered data fitting has been introduced to the approximation theory community by Schumaker [157] in 1976. Many different ideas and techniques have been developed since then. Among the most powerful approximation schemes are *splines*, including box splines [17] and splines on triangulations, and the meshfree *radial basis functions*, which are the subject of the extensive discussion in Chapter 3.

We remark that basic concepts and requirements of scattered data fitting are similar to those of *irregular sampling* in signal and image processing, although different techniques are used. Irregular sampling is mainly concerned with the *complete reconstruction* (rather than approximation) of a *band-limited* signal from its irregularly sampled values. To this end, irregular sampling is working with different tools from harmonic analysis in order to establish a rigorous time-frequency analysis in suitable function spaces. For a recent account on the state-of-the-art in irregular sampling we recommend the survey [3] by Aldroubi and Gröchenig, see also the earlier papers [68, 69, 70] by Feichtinger and Gröchenig.

Both scattered data fitting and irregular sampling are active research fields within their main application areas, approximation theory and information theory, with significant impact on many different disciplines in science and engineering. The focus in this work is more on scattered data fitting, with approaching the subject rather from the viewpoint of numerical approximation. We remark, however, that tools from scattered data approximation are also potentially useful for irregular sampling, and vice versa.

The modern concept of scattered data modelling has recently gained much attention in meshfree discretizations for numerically solving partial differential equations. In fact, many of the commonly used meshfree discretization techniques are point-based, where the enhanced flexibility of scattered data modelling techniques plays a key role, especially when it comes to modelling multiscale phenomena.

In fluid flow simulation, for instance, meshfree Lagrangian methods are used in order to numerically integrate the governing transport equations along *streamlines* of fluid particles. In this application, each particle is bearing specific physical information, such as concentration, saturation, or mass. Moreover, the time-dependent modelling process works with a discrete set $X \equiv X(t)$ of moving particles, which is subject to *dynamic* changes during the evolution of the flow.

This is in contrast to the *static* process of scattered data fitting, where the point cloud X is fixed. Nevertheless, the coupling of the particle model to continuous models requires approximation schemes from scattered data fitting.

Later in this work, adaptive meshfree methods for multiscale flow simulation are developed, where the coupling between the two models is accomplished. This is the subject of the extensive discussion in Chapter 6, where radial basis functions, as introduced in Chapter 3, play a key role.

1.2 Multiresolution Methods

Many different phenomena in natural sciences and engineering exhibit multiple levels of details. Among the wide range of examples are transport processes in fluid flow, where finer details of free turbulences may be due to irregular vortex motions or the evolution of shock fronts. Another example is the task of broadcasting signals and images, whose information is composed of waves of different frequency components.

The modelling of such *multiscale phenomena* is usually a computationally challenging task, which requires customized mathematical techniques, *multiresolution methods*, in order to represent the mathematical model at the relevant range of scales. The mathematical model needs to reflect the different levels of details by approximating the mathematical object on multiple different *scales*, ranging from a *coarse* representation at a *low resolution* to a *fine* representation at a *high resolution.*

But the computation and representation of the mathematical model at all possible scales of action is computationally too expensive, unless the multiscale nature of the underlying phenomenon is exploited in a fundamental way. Therefore, multiscale modelling requires efficient, robust, reliable and accurate numerical algorithms as well as flexible data structures and advanced techniques from computer programming for the modelling and visualization of such multiscale phenomena in computer simulations.

Effective multiresolution methods are essentially concerned with balancing the two conflicting requirements of low data size (computational efficiency) and high fidelity (approximation quality), where the goal is to keep the required computational costs at a given model resolution as small as possible. Therefore, multiscale modelling is usually concerned with information reduction, model simplification, and data compression. In fact, one important application of multiresolution methods is the progressive transmission of model objects, such as geometric objects and digital images, across an information channel, such as the internet.

Let us further discuss this important point by picking one concrete example from *geometric modelling.* In this particular application, a geometric object, for instance a solid body, may be represented in form of a triangular mesh. When it comes to fast rendering in a dynamic modelling process, it is clearly inefficient to display the whole triangular mesh, when the object is far from the viewer. In contrast to this, when the viewer zooms into local areas of the object, a more accurate representation of finer details is needed.

In this case, it is necessary to create a multiresolution model of the object beforehand by breaking down the original model into a hierarchical sequence of coarser and coarser versions, the coarsest one providing merely a rough impression of the original. Since the coarser detail levels contain much less data, and in particular fewer triangular faces, they can be displayed much faster than the finer detail levels. This, however, requires advanced techniques for the purpose of generating, storing, and manipulating geometric models. For a recent account on multiresolution methods in geometric modelling and their applications, see the tutorial book [99].

From the mathematical point of view, multiresolution modelling usually relies on the rigorous *multiresolution analysis* (MRA), which includes the theory of *wavelets*. In fact, wavelets have gained enormous popularity in applications such as signal and image processing. The mathematical theory of wavelets is well-understood when working with *regular* data sets. In this case, relatively simple mathematical objects, such as digital images are represented in form of rectangular grids of wavelet coefficients.

When working with scattered data, however, more general mathematical methods are necessary in order to establish a multiresolution analysis. In fact, a mathematical theory concerning multiresolution analysis in scattered data modelling is hardly developed. This is due to the inherent irregularity of scattered data that renders standard schemes, such as wavelet techniques and tensor product schemes, non-applicable.

Among the very few contributions in this research direction are wavelet-like schemes over nested and non-nested sequences of triangulations [75, 76]. For a comprehensive overview on wavelets and their applications in multiresolution modelling, we recommend the textbooks [28, 36], and the more recent tutorial papers [16, 31, 136].

Another way to generate multiresolution models is through *subdivision*. Subdivision schemes start off with a coarse mesh and generate finer and finer meshes according to some local scheme (a mask) and, provided the scheme is well-chosen, converge to a smooth surface. Much of the ground-breaking work on subdivision schemes in geometric modelling was done by Catmull and Clark [27], Doo and Sabin [45], and by Dyn, Gregory, and Levin [56]. We remark that the theory of subdivision schemes is intimately related to wavelet techniques, and thus also built on the useful concept of multiresolution analysis. For a very recent account on subdivision schemes in geometric modelling, we recommend the survey [57] by Dyn and Levin, the tutorial papers [52, 53] by Dyn, and [143, 144] by Sabin, as well as the short contribution [30] on applications of nonlinear subdivision to image processing.

There are basically two different concepts for designing multiresolution methods in scattered data modelling, both of which we utilize in this work. Firstly, the hierarchical decomposition of the model into several levels of detail leads to *multilevel methods*, some of whose features are introduced in Section 1.3. Pragmatic methods, such as *thinning algorithms* [40, 55, 74], were recently developed in order to generate a hierarchical sequence of scattered data by recursive point removals. These rather novel techniques show great potential in applications, as confirmed by their utility in digital image compression, see Section 4.6, and hierarchical surface visualization, see Section 5.4.

Secondly, a different approach for designing multiresolution methods is given by dynamic (one-level) modelling concepts in time-dependent processes, such as the evolution of fluid flow. In fact, much of this work is devoted to multiscale flow simulation by meshfree discretization techniques, see Chapter 6. This requires effective adaption rules, for the local refinement and coarsening of scattered nodes. Suitable such adaption rules are designed in Section 6.3. The underlying concept of *adaptive irregular sampling* is briefly explained in Section 1.4.

1.3 Multilevel Methods

Multilevel methods are concerned with generating a hierarchy $s_1, \ldots, s_L$ of approximations to a function f at various different resolutions, ranging from a coarse approximation to finer and finer approximations.

The coarsest approximation s_1 reflects the global trend of f, whereas finer details of f are captured by its subsequent approximations $s_2, \ldots, s_L$. Given a suitable norm $\|\cdot\|$, a decreasing sequence $\epsilon_1 > \epsilon_2 > \cdots > \epsilon_L > 0$ of model accuracies

$$\epsilon_\ell = \|f - s_\ell\|, \qquad \text{for } 1 \leq \ell \leq L,$$

is obtained by the different approximations s_ℓ to f. In multilevel modelling schemes, the transition between any coarse approximation, $s_{\ell-1}$, and the consecutive one, s_ℓ, is usually done by adding detail information, Δs_ℓ, in order to obtain the finer approximation

$$s_\ell = s_{\ell-1} + \Delta s_\ell, \quad 1 \leq \ell \leq L.$$

This is called the *refinement* of the model. Conversely, a transition in the opposite direction, from fine to coarse, is referred to as *coarsening* of the model.

Multilevel decompositions show great potential in applications, such as the fast rendering, editing and compression of geometric models, their transmission across the internet, computer animation, and scientific visualization in general. In these application fields, the execution of the transitions, mentioned above, need to be fast, since these operations typically need to be applied in real time during the dynamic modelling.

Multilevel methods in scattered data modelling usually require computing a hierarchy

$$X_1 \subset X_2 \subset \cdots \subset X_{L-1} \subset X_L = X \tag{1.1}$$

of nested sequences of the given point set X beforehand. This important task is referred to as *data analysis*. The data analysis is a very critical preprocess, since the performance (in terms of computational costs and approximation quality) of any multilevel approximation scheme, in the subsequent *data synthesis*, relies heavily on the *quality* of the hierarchy in (1.1).

Multilevel methods in scattered data modelling are dating back to [72], where the main ingredients of the multiresolution modelling are *thinning algorithms* (in the data analysis) and *radial basis functions* (in the data synthesis). Recent developments [96, 98] concerning multilevel approximation schemes are discussed in Chapter 5, whereas thinning algorithms are subject of Chapter 4, and radial basis functions are covered in Chapter 3.

1.4 Adaptive Irregular Sampling

In contrast to multilevel methods, adaptive irregular sampling leads to multiresolution models, where the multiple range of scales is incorporated in merely one single data level. The basic idea of adaptive irregular sampling is to use a large *sampling rate* (of large data density) in regions of high frequencies, whereas small sampling rates (of small data density) are considered in regions of low frequencies.

In this case, the resolution of the model varies adaptively in Ω, the domain of interest. More precisely, the model accuracy

$$\epsilon_U = \|f - s\|_U,$$

in a local neighbourhood $U \equiv U(x) \subset \Omega$ around any $x \in \Omega$ depends on the local behaviour of the model object f in the vicinity around x. This alternative multiresolution concept breaks down the global approximation problem into several local ones, and so the modelling is accomplished by using *local* approximation schemes. In this case, the terms *high resolution* versus *low resolution* depend on the different accuracy requirements in the different subregions of the domain Ω. High accuracy, and thus high resolution, in a local subregion $U \subset \Omega$ is necessary, whenever the mathematical object f is subject to strong variations in U. This is, for instance, required near *discontinuities* of f. In contrast to this, in subregions U where f is *smooth*, the model usually requires a much lower accuracy, and thus a lower resolution.

In scattered data modelling, adaptive irregular sampling leads to varying densities in the set $X \subset \Omega$ of *current* sample points. This concept is typically used for modelling time-dependent processes, such as in fluid flow simulation, where the points in X correspond to a set of moving fluid particles. In this case, the point set X and samples values $f|_X$ may be given *initially* at time $t = 0$, but the modelling usually relies on adaptive modifications of $X \equiv X(t)$ during the simulation. This in turn requires effective adaption rules for the dynamic modification of the points in X. Details on these and related aspects are treated in Chapter 6.

2 Algorithms and Data Structures

Much of the following discussion in this work relies on standard tools from computational geometry, approximation theory, and computer programming. For the reader's convenience, relevant material concerning required algorithms and data structures is composed in this chapter, which helps to keep this work widely self-contained. Moreover, notational preparations are done, and various useful auxiliary results are given.

This chapter first provides a general discussion on *triangulation methods* in Section 2.1, before *Delaunay triangulations* and their dual graphs, *Voronoi diagrams*, are introduced in the subsequent Sections 2.2 and 2.3. Then, Section 2.4 gives an introduction to *data-dependent triangulations*, which are useful methods for approximating bivariate functions from scattered data. Several available swapping criteria, which are required for the construction of data-dependent triangulations, are discussed in Section 2.4. Finally, Section 2.5 is concerned with the efficient implementation of priority queues by using the data structure *heap*, and Section 2.5 is devoted to domain decomposition by using the data structure *quadtree*.

2.1 Triangulation Methods

Triangulation methods are important ingredients in finite element methods (FEM), numerical approximation, computer-aided geometric design (CAGD), and various other application fields. For a discussion concerning triangulations in CAGD and in approximation theory, see [159, 160]. In these particular applications, a triangulation is typically used for partitioning a planar domain $\Omega \subset \mathbb{R}^2$. The partitioning is then in turn used for building approximation spaces, *splines on triangulations*, comprising piecewise polynomial functions.

Especially in bivariate scattered data modelling, splines on triangulations are powerful tools for the interpolation and approximation of surfaces. But before we proceed with this discussion, let us first give a formal definition for the term *triangulation.*

Definition 1. *A* `triangulation` *of a discrete point set $X = \{x_1, \ldots, x_N\}$ is a collection $\mathcal{T}_X = \{T\}_{T \in \mathcal{T}_X}$ of triangles in the plane, such that the following conditions hold.*

(a) *the vertex set of $\mathcal{T}_X$ is X;*
(b) *any pair of two distinct triangles in $\mathcal{T}_X$ intersect at most at one common vertex or along one common edge;*
(c) *the convex hull $[X]$ of X coincides with the area covered by the union of the triangles in $\mathcal{T}_X$.*

We say that a collection $\mathcal{T} = \{T\}_{T \in \mathcal{T}}$ of triangles is a `triangulation` *of a connected set $\Omega \subset \mathbb{R}^2$, iff property* **(b)** *holds and Ω is the union of the triangles in $\mathcal{T}$.*

Figure 2.1 shows a possible triangulation of a scattered point set of size $|X| = 20$. Note that for every discrete point set X comprising at least three points there always exists a triangulation $\mathcal{T}_X$. The triangulation $\mathcal{T}_X$ may, however, not be unique. In accordance with the above definition, a triangulation can also be regarded as a *planar graph*, whose *nodes* are the vertices of the triangles, and whose *edges* provide the connectivities in the graph [134]. Thus, the basic entities in a triangulation are *triangles*, *edges*, and *nodes* (*vertices*).

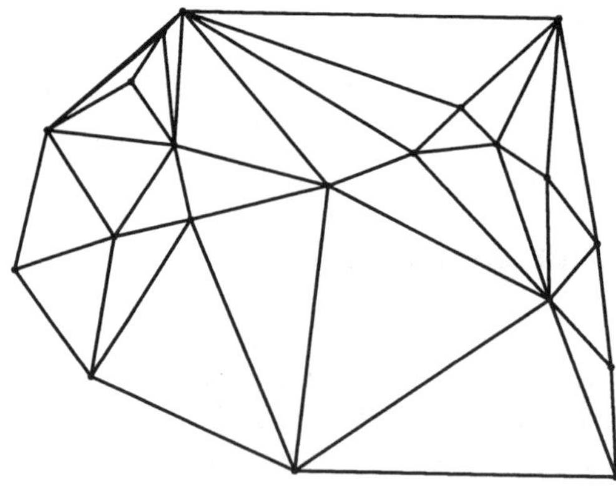

Fig. 2.1. A triangulation of a point set X of size $|X| = 20$.

By the classical *Euler formula* [134], we have, for any triangulation $\mathcal{T}$, the relationship

$$N_v - N_e + N_t = 1 \tag{2.1}$$

for its number N_t of triangles, N_e of edges, and $N = N_v$ of vertices. Moreover, with letting N^b the number of *boundary* points in $\mathcal{T}$, we obtain the formulae

$$N_t = 2N - N^b - 2$$
$$N_e = 3N - N^b - 3$$

for the number of triangles and edges in $\mathcal{T}$.

We associate with any triangulation $\mathcal{T}$ of Ω the *finite element space*

$$\mathcal{S}(\mathcal{T}) = \left\{ f \in C(\Omega) \, : \, f\big|_T \text{ linear; for all } T \in \mathcal{T} \right\}, \tag{2.2}$$

the linear function space comprising all continuous (real-valued) functions on Ω, whose restriction on any triangle in $\mathcal{T}$ is linear. In approximation theory, $\mathcal{S}(\mathcal{T})$ is usually called the space of *linear splines* on $\mathcal{T}$. Note that the dimension of $\mathcal{S}(\mathcal{T})$ is N. A cardinal basis for $\mathcal{S}(\mathcal{T})$ is given by the *Courant elements* $u_1, \ldots, u_N \in \mathcal{S}(\mathcal{T})$ satisfying $u_j(x_k) = \delta_{jk}$, for all $1 \leq j, k \leq N$, where x_k denotes the k-th vertex in $\mathcal{T}$.

Now suppose we are given real function values $f_1, \ldots, f_N$, sampled from an unknown function f at the vertex set $X = \{x_1, \ldots, x_N\}$ of $\mathcal{T}$. Then, the interpolation problem

$$s(x_k) = f_k, \qquad \text{for } 1 \leq k \leq N,$$

has a unique solution s in $\mathcal{S}(\mathcal{T})$, which can be expressed as

$$s = \sum_{j=1}^{N} f_j u_j.$$

The function s is said to be the *piecewise linear interpolant* to f over the triangulation $\mathcal{T}$. We make use of the notation $L(f, \mathcal{T}) = s$ in order to express the dependency of s on the triangulation $\mathcal{T}$ and the values of f at X.

We remark that $\mathcal{S}(\mathcal{T}) \equiv \mathcal{S}_1^0(\mathcal{T})$ is the simplest instance of the family of *bivariate splines*,

$$\mathcal{S}_q^r(\mathcal{T}) = \left\{ f \in C^r(\Omega) \, : \, f\big|_T \in \mathcal{P}_{q+1}^2 \text{ for all } T \in \mathcal{T} \right\}, \qquad \text{for } q > r \geq 0,$$

containing all functions on Ω which are globally of smoothness C^r, and whose restriction on any $T \in \mathcal{T}$ is an element of $\mathcal{P}_{q+1}^2$, the linear space of all bivariate polynomials of order at most $q+1$ (and thus of degree at most q). Hence, $\mathcal{S}_q^r(\mathcal{T})$ consists of piecewise polynomial functions on Ω.

Bivariate splines provide powerful and stable methods for scattered data fitting. These classical techniques are, however, not in the focus of this work. Therefore, we do not dwell on explaining further details on this wide research field. Instead of this, we prefer to recommend the recent tutorial [178] for an up-to-date account on the theory of bivariate splines and their applications.

2.2 Delaunay Triangulations

Apart from trivial cases, there are many different triangulations of a given planar point set X. Let us first look at the simple example, where the set

$X = \{x_1, x_2, x_3, x_4\}$ comprises merely four points. Furthermore, we assume that the four points in X are the vertices of a strictly convex quadrilateral $\mathcal{Q}$, i.e., $\mathcal{Q}$ is convex and no three points of its vertex set X are co-linear, see Figure 2.2. For brevity, we use from now on the term *convex quadrilateral* rather than *strictly convex quadrilateral.*

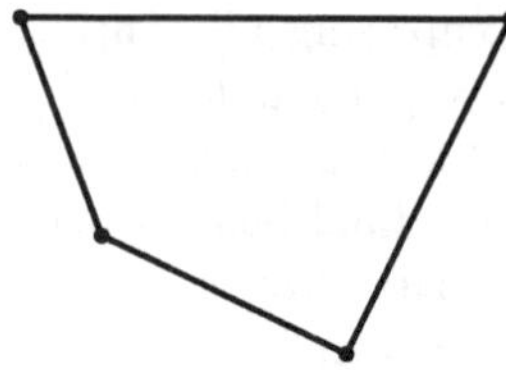

Fig. 2.2. A strictly convex quadrilateral $\mathcal{Q}$.

There are obviously two different ways for splitting $\mathcal{Q}$ into two triangles, yielding two different triangulations $\mathcal{T}$ and $\tilde{\mathcal{T}}$ of X, as shown in Figure 2.3. We denote by e the diagonal edge in $\mathcal{T}$, and by $\tilde{e}$ the diagonal edge in $\tilde{\mathcal{T}}$.

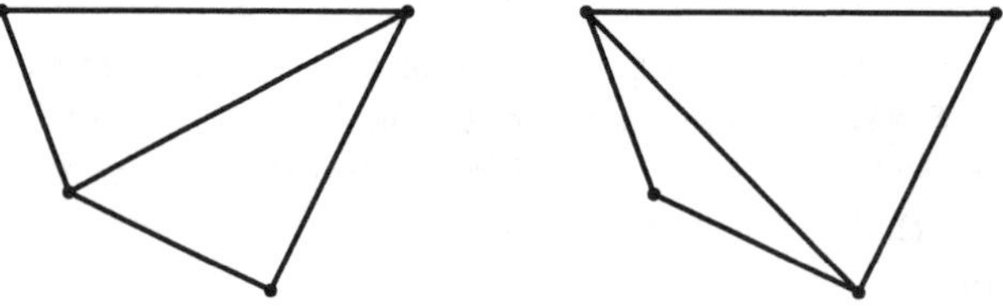

Fig. 2.3. Two triangulations of a convex quadrilateral, $\mathcal{T}$ (left) and $\tilde{\mathcal{T}}$ (right).

When it comes to choosing between $\mathcal{T}$ and $\tilde{\mathcal{T}}$, there are several good reasons which would favour $\mathcal{T}$, the left one in Figure 2.3. These mathematical reasons rely on error estimates from finite element methods on triangulations, depending on the largest diameter in the triangulation [18, 81, 158]. This suggests that one should avoid long thin triangles, as occurring in $\tilde{\mathcal{T}}$, the right one in Figure 2.3. We remark at this point that there are also good mathematical reasons in favour of using long thin triangles. Details on this are explained in Section 2.4.

But if we wish to avoid long and thin triangles, this requires keeping the smallest angle in the triangulation as large as possible. According to this criterion, a triangulation $\mathcal{T}$ is *better* than $\tilde{\mathcal{T}}$, $\mathcal{T} > \tilde{\mathcal{T}}$, iff the minimal angle

$$\alpha(\mathcal{T}) = \min \{\alpha_{\min}(T) : T \in \mathcal{T}\} \tag{2.3}$$

in $\mathcal{T}$ is larger than the minimal angle $\alpha(\tilde{\mathcal{T}})$ in $\tilde{\mathcal{T}}$, where $\alpha_{\min}(T)$ in (2.3) denotes the minimal angle of a triangle T. In the situation of Figure 2.3, we have $\alpha(\mathcal{T}) = 27.47°$ and $\alpha(\tilde{\mathcal{T}}) = 19.98°$. Therefore, $\mathcal{T}$ is better than $\tilde{\mathcal{T}}$.

In situations where the four points in X are *co-circular*, i.e., they lie on the circumference of a circle, it is easy to show that neither of the two possible triangulations of $\mathcal{Q}$, $\mathcal{T}$ or $\tilde{\mathcal{T}}$, is better than the other. In this case, they are said to be *equally good*, $\mathcal{T} = \tilde{\mathcal{T}}$.

If, however, the points in X are not co-circular, then one triangulation, either $\mathcal{T}$ or $\tilde{\mathcal{T}}$, must be better than the other. This is confirmed by the following lemma due to Lawson [109], which provides an alternative criterion, the *Delaunay criterion*, for checking this.

Lemma 1. *Let $\mathcal{Q}$ be a convex quadrilateral whose four vertices are not co-circular. Furthermore, let $\mathcal{T}, \tilde{\mathcal{T}}$ denote the two possible triangulations of $\mathcal{Q}$. Then, $\mathcal{T}$ is better than $\tilde{\mathcal{T}}$, iff for each of the two triangles in $\mathcal{T}$ its circumcircle does not contain any point from the vertex set X of $\mathcal{Q}$ in its interior. Otherwise, $\tilde{\mathcal{T}}$ is better than $\mathcal{T}$.*

If $\mathcal{T}$ is better or equally as good as $\tilde{\mathcal{T}}$, $\mathcal{T} \geq \tilde{\mathcal{T}}$, then the quadrilateral $\mathcal{Q}$ is said to be *optimally triangulated* by $\mathcal{T}$. In this case, we also say that $\mathcal{T}$ is an *optimal triangulation* for $\mathcal{Q}$, or, more briefly, we say that the edge e in $\mathcal{T}$ is *optimal*. In the following discussion, it is convenient to say that any interior edge, whose two adjacent triangles form a *non-convex* quadrilateral, is optimal.

Note that any convex quadrilateral $\mathcal{Q}$, which is not optimally triangulated, can be transferred into an optimally triangulated one by an *edge swap*. In this case, the current triangulation, say $\mathcal{T}$, of $\mathcal{Q}$ is replaced by the other possible triangulation of $\mathcal{Q}$, $\tilde{\mathcal{T}}$. In other words, the current (non-optimal) edge e on the diagonal of $\mathcal{Q}$ is swapped for the edge $\tilde{e}$ on the opposite diagonal of $\mathcal{Q}$. The edge $\tilde{e}$, and so the triangulation $\tilde{\mathcal{T}}$ of $\mathcal{Q}$, is in this case optimal.

Now let us return to the general situation, where the triangulation may comprise arbitrarily many vertices. After the above discussion we are in a position to introduce Delaunay triangulations [134].

Definition 2. *The* `Delaunay triangulation` *$\mathcal{D}_X$ of a discrete planar point set X is a triangulation of X, such that the circumcircle for each of its triangles does not contain any point from X in its interior.*

Figure 2.4 shows an example for the Delaunay triangulation of a point set of size $|X| = 20$. We remark that the Delaunay triangulation $\mathcal{D}_X$ of X is unique, provided that no four points in X are co-circular [134]. Note that by the above definition, any interior edge in a Delaunay triangulation is optimal, i.e., any convex quadrilateral $\mathcal{Q}$ in $\mathcal{D}_X$ satisfies the Delaunay criterion, and so $\mathcal{Q}$ is optimally triangulated.

Moreover, one can show that the Delaunay triangulation $\mathcal{D}_X$ of a point set X maximizes the minimal angle among all possible triangulations of X, see [159]. In this sense, Delaunay triangulations are (globally) *optimal* triangulations.

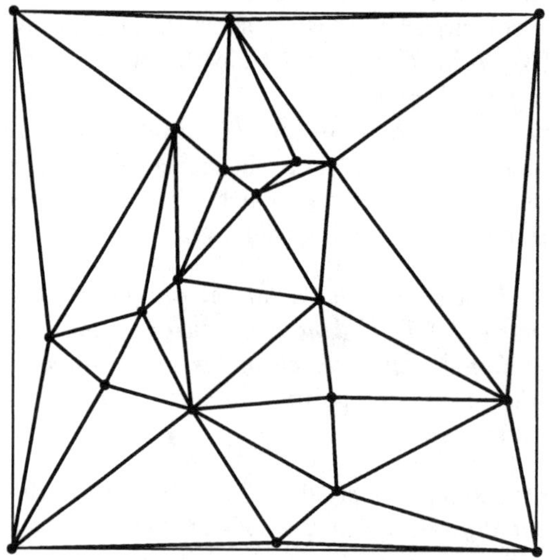

Fig. 2.4. The Delaunay triangulation of a point set comprising 20 points.

Considerable effort has gone into the design of algorithms for the construction of Delaunay triangulations. There are basically three different approaches, yielding the three Algorithms 1, 2 and 3 below, each of which relies on the Delaunay criterion in Lawson's Lemma 1. In fact, the first Algorithm 1 is based on the *local optimization procedure* (LOP) due to Lawson, Algorithm 4 in Section 2.4.

Algorithm 1 (Post optimize).
Construct an initial triangulation. Then iteratively go through the list of convex quadrilaterals and make edge swaps for non-optimal edges.

Algorithm 2 (Iteratively build).
Start with one triangle (with vertices in X) and add one point from X at a time, making sure that at each step every convex quadrilateral in the current triangulation is optimally triangulated.

Algorithm 3 (Divide and conquer).
Recursively divide the data points up into pieces, find a Delaunay triangulation for each piece, and then merge these triangulations.

It can be shown that every triangulation algorithm requires at least $\mathcal{O}(N \log(N))$ operations, where N is the number of points in the set to be triangulated [159]. As to the computational costs required for the above three algorithms, the Algorithm 3 (divide and conquer) has optimal complexity, $\mathcal{O}(N \log(N))$, whereas the other two, Algorithm 1 and 2 cost $\mathcal{O}(N^2)$ in the worst case scenario, but $\mathcal{O}(N \log(N))$ in average.

2.3 Voronoi Diagrams

For any Delaunay triangulation, there is a dual graph, called the *Voronoi diagram*. Before we explain this duality relation, which has been observed by Delaunay [39] in 1934, let us make a few general remarks concerning Voronoi diagrams. Voronoi diagrams (and thus Delaunay triangulations) are well-suited for solving *closest point problems* on a given point set X, such as the question: on given point y, which point in X is closest to y?

In fact, the Voronoi diagram of a point set X consist of *Voronoi tiles*, each of which being associated with one point x in X. The Voronoi tile of x contains all points which are at least as close to x as to any other point in X. Voronoi diagrams, also referred to as *Dirichlet, Thiessen*, or *Voronoi tessellations*, have a variety of important applications in biology, computer-aided design, geography, and many other fields.

For the subsequent discussion in this work, it is convenient to formulate the duality relation between Voronoi diagrams and Delaunay triangulations in general dimension d. Suppose $X \subset \mathbb{R}^d$ is a finite point set in $\mathbb{R}^d$. Then, for any $x \in X$, the point set

$$V_X(x) = \left\{ y \in \mathbb{R}^d \; : \; \|y - x\| = \min_{\bar{x} \in X} \|y - \bar{x}\| \right\} \subset \mathbb{R}^d$$

is said to be the *Voronoi tile* of x, and the collection $\{V_X(x)\}_{x \in X}$ of Voronoi tiles is called the *Voronoi diagram* of X. Here, $\| \cdot \|$ is the Euclidean norm on $\mathbb{R}^d$. Note that the Voronoi diagram yields a partitioning of the Euclidean space,

$$\mathbb{R}^d = \bigcup_{x \in X} V_X(x).$$

Moreover, each Voronoi tile $V_X(x)$ is a non-empty, closed and convex polyhedron, whose vertices are called *Voronoi points*. Two different Voronoi tiles $V_X(x)$ and $V_X(y)$ are either disjoint or they share a common face, in which case the points $x \in X$ and $y \in X$ are said to be *Voronoi neighbours*.

By connecting all possible Voronoi neighbours, we obtain a graph whose vertex set is X. This graph defines a *simplicial decomposition* $\mathcal{D}_X$ of the convex hull $[X]$, provided that no $d+2$ points in X are *co-spherical*. The latter means that no $d+2$ points in X lie on the $(d-1)$-dimensional surface of a sphere $S \subset \mathbb{R}^d$. Let us from now on assume this property, which helps us to omit lengthy and tedious but inconsequential technical details concerning degenerate cases.

The simplicial decomposition $\mathcal{D}_X$ of X is said to be the *Delaunay triangulation* of X. In the special case of two dimensions, $\mathcal{D}_X$ is the planar Delaunay triangulation, as introduced in the previous Section 2.2. Thus, there are two alternative (and equivalent) definitions for a Delaunay triangulation in two dimensions, one given by Definition 2, and the other by regarding the dual graph of the Voronoi diagram. Figure 2.5 shows a Voronoi diagram, the dual one of the Delaunay triangulation in Figure 2.4.

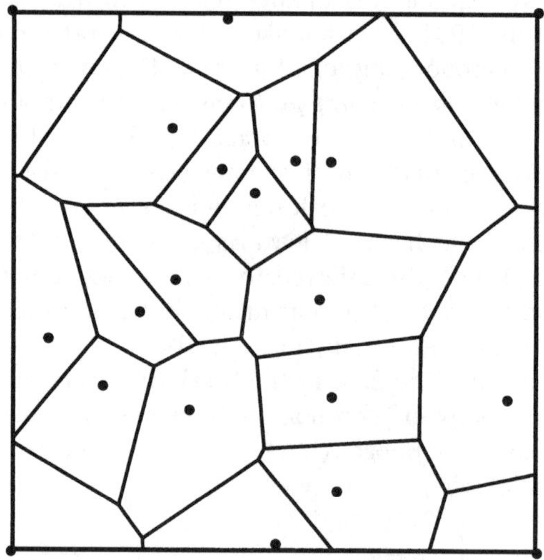

Fig. 2.5. The Voronoi diagram of a point set comprising 20 points.

Likewise, an alternative and equivalent definition for the Delaunay triangulation in arbitrary dimension can be obtained by the generalization of the Delaunay criterion in the previous section.

Definition 3. *The* `Delaunay triangulation` $\mathcal{D}_X$ *of a discrete point set* $X \subset \mathbb{R}^d$ *is a simplicial decomposition, whose vertex set is* X *and such that the circumsphere of each of its simplices does not contain any point from* X *in its interior.*

The triangulation $\mathcal{D}_X$ of X is unique (provided that no $d+2$ points in X are *co-spherical*). Moreover, each center point of a circumsphere in a simplex of $\mathcal{D}_X$ is a Voronoi point. Note that the Euclidean *distance function* d_X associated with X, and defined by

$$d_X(y) = \min_{x \in X} \|y - x\|, \tag{2.4}$$

has local maxima at the Voronoi points, where $\|\cdot\|$ is the Euclidean norm on $\mathbb{R}^d$.

We close this section by remarking that for any $x \in X$, all of its nearest points from $X \setminus x$ are Voronoi neighbours of x. In other words, each vertex $x \in X$ in the Delaunay triangulation $\mathcal{D}_X$ of X is connected with all of its nearest vertices. In particular, a closest point pair $x, y \in X$ satisfying

$$\|x - y\| = d_{X \setminus x}(x) = d_{X \setminus y}(y)$$

is connected by an edge in $\mathcal{D}_X$. Hence, the task of localizing such a closest pair of points boils down to locating the shortest edge in $\mathcal{D}_X$. We make advantage of this important property later in this work.

2.4 Data-Dependent Triangulations

Now let us return to planar triangulations. It has for long time been widely accepted that when it comes to fitting bivariate scattered data by splines on triangulations, the Delaunay triangulation of the data points is the best one to use, because it avoids long thin triangles. The argumentation is based on available error bounds for finite element approximations, depending on the largest diameter in the triangulation [18, 81, 158].

For interpolation by *linear* splines, however, it was first numerically shown in [58, 59, 60] by Dyn, Levin and Rippa, and later proven in [140] by Rippa, that long and thin triangles can improve the approximation quality of the resulting piecewise linear interpolant. This is also confirmed in [137] for C^1 piecewise cubic elements on triangulations.

The approach in [58, 59, 60] (see also the review [61]) is based on the concept of *data-dependent triangulations*, triangulations whose construction essentially depends on the function values of the underlying function f to be approximated. Note that this is in contrast to Delaunay triangulations, whose construction depends only on the locations of the points in X, but not on the function values of f at X, see the Algorithms 1, 2, and 3 in Section 2.2.

Starting point for the construction of any data-dependent triangulation (DDT) is the selection of one specific optimality criterion for convex quadrilaterals, like the Delaunay criterion in Delaunay triangulations. In fact, the construction of data-dependent triangulations is done by using the Algorithm 1 in Section 2.2, but by working with different criteria (other than the Delaunay criterion) for swapping edges.

Any such (data-dependent) swapping criterion assigns a *cost value* $c(e)$ to each interior edge e of a triangulation. This value can be viewed as the *energy* of the edge e. When it comes to choosing between the two different triangulations $\mathcal{T}$ and $\tilde{\mathcal{T}}$ of any convex quadrilateral $\mathcal{Q}$ (see Figure 2.3), the cost values $c(e)$ and $c(\tilde{e})$ of the two interior edges $e, \tilde{e}$ of $\mathcal{T}$ and $\tilde{\mathcal{T}}$ are compared. In the situation of Delaunay triangulations, the cost function may be given by $c(e) = 1/\alpha(\mathcal{T})$, the reciprocal value of the minimal angle in $\mathcal{T}$.

If $c(e) < c(\tilde{e})$, then $\mathcal{T}$ is better than $\tilde{\mathcal{T}}$, $\mathcal{T} > \tilde{\mathcal{T}}$. Moreover, if $c(e) = c(\tilde{e})$, then the two triangulations $\mathcal{T}$ and $\tilde{\mathcal{T}}$ are equally good, $\mathcal{T} = \tilde{\mathcal{T}}$. Finally, if

$\mathcal{T}$ is better than or as good as $\tilde{\mathcal{T}}$, $\mathcal{T} \geq \tilde{\mathcal{T}}$, then the edge e is said to be *optimal.* Vice versa, if $\tilde{\mathcal{T}} \geq \mathcal{T}$, then the edge $\tilde{e}$ is said to be optimal. Again, any interior edge of a *non-convex* quadrilateral is said to be optimal.

Now the overall *quality* of any triangulation $\mathcal{T}$ can be measured by using the cost values of its interior edges, $e_1, \ldots, e_n$, which yields the *cost vector* $c(\mathcal{T}) = (c(e_1), \ldots, c(e_n))^T \in \mathbb{R}^n$ of the triangulation $\mathcal{T}$. This allows us to compare two different triangulations (of one fixed point set X), $\mathcal{T}_X$ and $\tilde{\mathcal{T}}_X$, by using their cost vectors $c(\mathcal{T}_X)$ and $c(\tilde{\mathcal{T}}_X)$. Recall from Euler's formula (2.1) that the number of interior edges in $\mathcal{T}_X$ and $\tilde{\mathcal{T}}_X$, and thus the length of their cost vectors, are equal. Now if $\|c(\mathcal{T}_X)\| \leq \|c(\tilde{\mathcal{T}}_X)\|$, for some norm $\|\cdot\|$ on $\mathbb{R}^n$, then the triangulation $\mathcal{T}_X$ is said to be better than or as good as $\tilde{\mathcal{T}}_X$, $\mathcal{T}_X \geq \tilde{\mathcal{T}}_X$. This leads us to the following definition for optimal triangulations.

Definition 4. *A triangulation $\mathcal{T}_X^*$ of a finite point set X is said to be* `optimal` *, iff $\mathcal{T}_X^* \geq \mathcal{T}_X$ holds for every triangulation $\mathcal{T}_X$ of X.*

Note that for any finite point set X there is always one optimal triangulation $\mathcal{T}^*$ of X, which may possibly be not unique though. Although it would be desirable to construct, on any given point set X, an optimal triangulation $\mathcal{T}_X^*$, the computational costs for the required global optimization is in general too expensive. Therefore, we prefer to work with *locally optimal* triangulations, whose construction is much cheaper.

Definition 5. *A triangulation is said to be* `locally optimal` *, iff all of its interior edges are optimal.*

Note that every optimal triangulation is also locally optimal. Indeed, suppose $\mathcal{T}^*$ is an optimal triangulation, which is not locally optimal. Then there is at least one interior edge e in $\mathcal{T}^*$, which is not optimal. Thus, by swapping e, the triangulation $\mathcal{T}^*$ is transferred into a triangulation $\tilde{\mathcal{T}}$, such that $c(\tilde{\mathcal{T}}) < c(\mathcal{T}^*)$, which is in contradiction to the optimality of $\mathcal{T}^*$.

Now let us turn to the construction of locally optimal triangulations. Having computed any initial triangulation $\mathcal{T}_0$ of a given point set X, e.g. its Delaunay triangulation $\mathcal{T}_0 \equiv \mathcal{D}_X$, a locally optimal (data-dependent) triangulation can be constructed by iteratively swapping non-optimal edges. This is the aforementioned *local optimization procedure* (LOP) due to Lawson [110].

Algorithm 4 (Local Optimization Procedure).

INPUT: *A finite point set $X \subset \mathbb{R}^2$, and function values $f\big|_X$.*

(1) *Construct an initial triangulation $\mathcal{T}_0$ of X, and let $\mathcal{T} = \mathcal{T}_0$.*
(2) WHILE ($\mathcal{T}$ *is not locally optimal*) **DO**
  (2a) *Locate a non-optimal edge e in $\mathcal{T}$;*
  (2b) *Swap the edge e, and so obtain the modified triangulation $\tilde{\mathcal{T}}$;*
  (2c) *Let $\mathcal{T} = \tilde{\mathcal{T}}$.*

OUTPUT: *A locally optimal triangulation $\mathcal{T}$.*

Note that this LOP terminates after finitely many edge swaps. Indeed, since the input point set X is assumed to be finite, and each edge swap strictly reduces the non-negative energy $c(\mathcal{T})$ of the current triangulation $\mathcal{T}$, this process must terminate after finitely many steps.

In the remainder of this section, we discuss various useful swapping criteria for the construction of data-dependent triangulations, although we do not utilize all of them in the following of this work.

One important point for the construction of a DDT is *shape-preservation.* If, for instance, f is convex, the resulting piecewise linear interpolation $L(f, \mathcal{T})$ over the resulting triangulation $\mathcal{T}$ should preserve this property, i.e., $L(f, \mathcal{T})$ should also be convex, in which case $\mathcal{T}$ is said to be a *convex triangulation.* The desired convexity property for $\mathcal{T}$ can always be achieved by using the following swapping criterion, suggested by Mulansky [125], in combination with Lawson's local optimization procedure, Algorithm 4.

Criterion 1 (Convexity).
Let $\mathcal{Q}$ be a convex quadrilateral, and let $\mathcal{T}$ and $\tilde{\mathcal{T}}$ denote the two different triangulations of $\mathcal{Q}$. Then, the diagonal edge e of $\mathcal{T}$ is optimal, iff the piecewise linear interpolant $L(f, \mathcal{T})$ is convex.

Note that for convex data, there is always at least one triangulation of a convex quadrilateral $\mathcal{Q}$, $\mathcal{T}$ or $\tilde{\mathcal{T}}$, which is convex. In case of constant data, both triangulations are convex. In this case, $L(f, \mathcal{T}) \equiv L(f, \tilde{\mathcal{T}})$, and so they are equally good, $\mathcal{T} = \tilde{\mathcal{T}}$. Mulansky [125] gave also an expression for a corresponding cost functional, which shows that for *convex* f the LOP outputs after finitely many swaps a *convex* triangulation $\mathcal{T}_X^*$ satisfying

$$L(f, \mathcal{T}_X^*) \leq L(f, \mathcal{T}_X),$$

pointwise on $\Omega = [X]$, for all triangulations $\mathcal{T}_X$ of X. Moreover, $L(f, \mathcal{T}_X^*)$ is in this case the *best approximation* of f among all piecewise linear functions over the triangulations of X, i.e., the inequality

$$|f(x) - L(f, \mathcal{T}_X^*)(x)| \leq |f(x) - L(f, \mathcal{T}_X)(x)|, \qquad \text{for all } x \in \Omega, \tag{2.5}$$

holds for every triangulation $\mathcal{T}_X$ of X. So in this sense, convex triangulations are (globally) *optimal.*

Several other useful data-dependent swapping criteria are proposed in [58, 59, 60] by Dyn, Levin and Rippa. Each of these swapping criteria is well-motivated by using reasonable assumptions on **(a)** the smoothness of the underlying function f, **(b)** possible variational properties of f, and **(c)** specific criteria based on the resulting approximation error. This yields three different classes of swapping criteria for data-dependent triangulations.

(a) Nearly C^1 criteria;
(b) Variational criteria;
(c) Minimal error criteria.

Various useful swapping criteria are discussed in the remainder of this section.

Nearly C^1 Criteria (NC1). For a variety of different test cases it is shown in [58, 59, 60] that data-dependent triangulations, which are producing long and thin triangles can significantly improve the approximation quality of the resulting piecewise linear interpolation. In particular, Rippa proves in [140] that long thin triangles are well-suited for approximating a *smooth* surface f with a preferred direction, i.e., f has large second-order directional derivatives in one direction, compared with other directions. In this case, the long side of the triangles should point into the directions of small curvature.

Since f is *smooth*, and so f does not oscillate too much between the data points, the construction of data-dependent swapping criteria which prefer *smooth* transitions between triangles across common edges is recommended. In this case, a suitable swapping criterion measures the discontinuity of first order derivatives across interior edges. Among several criteria which were proposed in [58, 59, 60], two turned out to perform particularly well in terms of their resulting approximation behaviour, *angle between normals* (ABN) and *jump of normal derivatives* (JND).

In order to define the required cost functions of ABN and JND, we need to make some notational preparations. Let e denote an interior edge in a convex quadrilateral $\mathcal{Q}$, and let T_1 and T_2 denote the two triangles in the resulting triangulation $\mathcal{T}$ of $\mathcal{Q}$, i.e., T_1 and T_2 share the common edge e. Moreover, let the piecewise linear function $L(f, \mathcal{T})$ of f on $\mathcal{T}$ be composed by the two linear functions

$$P_1(x, y) = a_1 x + b_1 y + c_1 \quad \text{and} \quad P_2(x, y) = a_2 x + b_2 y + c_2,$$

where P_1 is corresponding to T_1, and P_2 is corresponding to T_2.

Criterion 2 (Angle Between Normals, ABN).
The angle θ_e between the two normal vectors

$$n_i = \frac{1}{\sqrt{a_i^2 + b_i^2 + 1}} \cdot (a_i, b_i, -1)^T, \quad i = 1, 2,$$

of the two planes P_1 and P_2 is given by the expression

$$\cos(\theta_e) = \frac{n_1^T \cdot n_2}{\|n_1\|_2 \cdot \|n_2\|_2} = \frac{a_1 a_2 + b_1 b_2 + 1}{\sqrt{(a_1^2 + b_1^2 + 1)(a_2^2 + b_2^2 + 1)}}.$$

The ABN cost function is then given by $c(e) = \theta_e$.

Note that the criterion ABN basically measures the angle between the normals of two planes in $\mathbb{R}^3$. Therefore, the criterion ABN is also well-defined for surface triangulations of point sets in $\mathbb{R}^3$. In this case, the piecewise linear surface is no longer a bivariate function. Such surfaces do typically appear in solid modelling, for instance.

Criterion 3 (Jump of Normal Derivatives, JND).
The jump in the normal derivatives of P_1 and P_2 across the edge e is measured by

$$j(e) = |n_x(a_1 - a_2) + n_y(b_1 - b_2)|,$$

where $n = (n_x, n_y)^T \in \mathbb{R}^2$ is a unit vector that is orthogonal to the direction of the edge e. In this case, the JND cost function is given by $c(e) = j(e)$.

Variational Criteria. Suppose that the unknown function f can be characterized by a variational property of the form

$$I(f) = \min_{g \in \mathcal{F}} I(g),$$

where $I : \mathcal{F} \to \mathbb{R}$ is an *energy functional* on a suitable function space $\mathcal{F}$, comprising f and all piecewise linear functions over triangulations. In this case, it makes sense to select a criterion which yields a triangulation $\mathcal{T}$ of X, whose corresponding piecewise linear approximation $L(f, \mathcal{T}) \in \mathcal{F}$ to f minimizes the energy $I(L(f, \mathcal{T}))$ among all other triangulations of X.

One example for such a variational criterion is given by the *minimal roughness criterion*, whose energy functional is the Sobolev semi-norm

$$I(f) = |f|_{\Omega,1}$$

on $\Omega = [X]$, and so $\mathcal{F}$ is the Sobolev space $\mathcal{H}_1(\Omega)$. For any triangulation $\mathcal{T}$ of X, the roughness of $s = L(f, \mathcal{T}) \in \mathcal{S}(\mathcal{T})$ is in this case given by

$$|s|^2_{\Omega,1} = \sum_{T \in \mathcal{T}} |s|^2_{T,1},$$

where for any triangle $T \in \mathcal{T}$, and with letting $x = (\xi, \eta) \in \Omega$,

$$|s|^2_{T,1} = \int_T \left(s_\xi^2 + s_\eta^2\right) \, d\xi \, d\eta.$$

Interestingly enough, due to Rippa [139] the (data-independent) Delaunay triangulation $\mathcal{D}_X$ minimizes the roughness among all triangulations of X, i.e.,

$$|L(f, \mathcal{D}_X)|_{\Omega,1} \le |L(f, \mathcal{T}_X)|_{\Omega,1} \tag{2.6}$$

holds for all triangulations $\mathcal{T}_X$ of X. Therefore, the Delaunay triangulation is (globally) *optimal* with respect to this variational criterion.

It is well-known (see e.g. [141]) that the property (2.6) implies that

$$|f - L(f, \mathcal{D}_X)|_{\Omega,1} \le |f - L(f, \mathcal{T}_X)|_{\Omega,1} \tag{2.7}$$

holds for any triangulation $\mathcal{T}_X$ of X, provided that f is a *harmonic function*, i.e., $\Delta f = 0$, where

$$\Delta = \frac{\partial^2}{\partial \xi^2} + \frac{\partial^2}{\partial \eta^2}$$

denotes the *Laplace operator*. So in this case, the piecewise linear interpolant $L(f, \mathcal{D}_X)$ over the Delaunay triangulation $\mathcal{D}_X$ of X is the best approximation of f among all piecewise linear interpolants $L(f, \mathcal{T}_X)$.

Minimal Error Criteria. Given a normed linear function space $\mathcal{F}$ comprising all piecewise linear functions over triangulations, and a fixed $f \in \mathcal{F}$, the approximation quality of any triangulation $\mathcal{T}$ can be measured by

$$\epsilon(f, \mathcal{T}) = \|f - L(f, \mathcal{T})\|.$$

In this case, it is desirable to find a best approximation $L(f, \mathcal{T}^*)$ satisfying

$$\epsilon(f, \mathcal{T}^*) \leq \epsilon(f, \mathcal{T})$$

for all triangulations $\mathcal{T}$ of X, in which case the triangulation $\mathcal{T}^*$ is optimal. Examples for optimal triangulations of such kind, output by the local optimization procedure of Algorithm 4, are convex triangulations, minimizing the error in (2.5), and Delaunay triangulations, minimizing the error in (2.7).

2.5 Heaps and Priority Queues

A *heap* is a standard data structure which supports the efficient sorting of data objects. The resulting sorting algorithm is called *heapsort*, dating back to Williams [173]. Heapsort sorts the objects by using the values of their *keys*. The key of an object assigns a unique *significance value* to the object. The task of the sorting algorithm is to order the objects by their significances. The ordering can either be taken in increasing or decreasing order. For the moment of the discussion in this section we restrict ourselves to decreasing order, i.e., the most significant object comes first.

A heap is an array $A = A(1, \ldots, n)$ of data objects $A(i)$, $i = 1, \ldots, n$, that can be viewed as a binary tree, see Figure 2.6. Each node of the tree corresponds to a data object $A(i)$ of the array, whose significance value $\sigma(A(i))$ is stored in $A(i)$. The *root* of the tree is $A(1)$. The two *children* of the root are the nodes $A(2)$ and $A(3)$, and the root is the *parent* of these two nodes. More general, for any admissible index i the two children of the node $A(i)$ are $A(2i)$ and $A(2i + 1)$, where $A(2i)$ is called the *left child* and $A(2i + 1)$ is called the *right child* of $A(i)$. Conversely, the parent of any node $A(i)$ is $A(\lfloor i/2 \rfloor)$. Note that every node in the heap, except for the root, has a unique parent node. But not every node has two children. In fact, a node may have either two children, one child, or no children. A node without any children is called a *leaf*. In the situation of Figure 2.6, the nodes $A(5), \ldots, A(9)$ are the leaves of the heap. The tree is completely filled on all levels except possibly the lowest. But the lowest level is filled from left to right, so that there are no gaps in the object array A. Hence the *height* of the tree, being the number of levels between the root and the leaves, is $\log_2(n) \sim \log(n)$.

The nodes in a heap need to be organized such that for each node $A(i)$ its significance value is less than or equal the significance value of its parent $A(\lfloor i/2 \rfloor)$, i.e.,

$$\sigma(A(i)) \leq \sigma(A(\lfloor i/2 \rfloor)), \qquad \text{for } i > 1,$$

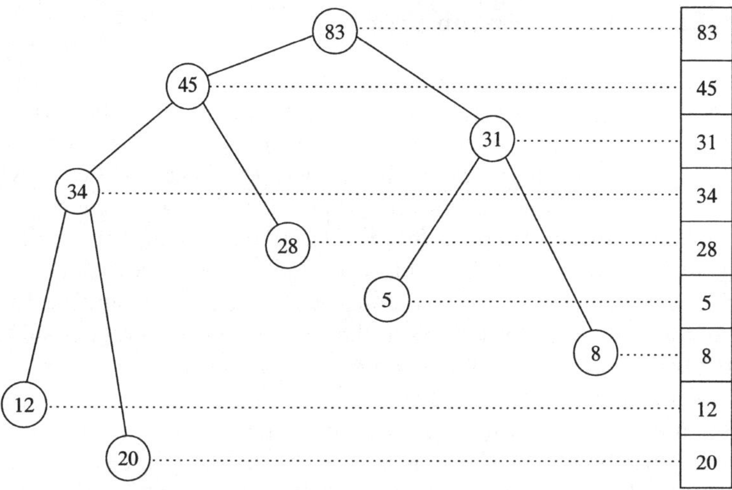

Fig. 2.6. A heap A = (83,45,31,34,28,5,8,12,20) of length $n = 9$ and height $h = 4$.

see Figure 2.6. This property is referred to as the *heap condition.* Therefore, according to this condition, the root $A(1)$ is the most significant node. Moreover, every node $A(i)$ in the heap has a larger significance than any node in the subtree rooted at $A(i)$. Note that the ordering of the data objects in the heap allows us to quickly retrieve the most significant object, $A(1)$, of the whole data set. Therefore, the heap data structure is also well-suited for the efficient implementation of a *priority queue.*

In this work, we are mainly interested in using heaps for the implementation of priority queues rather than sorting data objects, although these two tasks are strongly related to each other. But let us briefly explain our motivation for using heaps. In the situation of forthcoming applications, the data objects are points, and so the whole data set corresponds to a point cloud. We design algorithms which dynamically modify the point cloud by the removal and insertion of points, one at a time. By each such modification, the significance values of some of the points are updated. At any time step, we are interested in efficiently locating the most significant (least significant) point from the point cloud. To this end, we store the points in a heap, so that according to the heap condition the most significant (least significant) point is the root. The dynamic modification of the point cloud requires efficiently updating the heap with maintaining the heap condition. In the following of this section, the suitability of the heap data structure for our purposes is shown. More precisely, we discuss the construction of a heap, explain elementary operations on heaps, such as the removal and insertion of nodes, and we analyze the required computational costs, respectively.

2.5.1 Maintaining the Heap Property

Starting point of the following discussion is the important routine called `heapify`, which serves to maintain the heap property. On given array A and index i, the routine assumes that the two subtrees rooted at the two children $A(2i)$ and $A(2i+1)$ of the node $A(i)$ satisfy the heap condition, whereas $A(i)$ may have a smaller significance than its children, in which case the node $A(i)$ violates the heap condition. The function `heapify`, displayed in Algorithm 5, manipulates the subtree rooted at $A(i)$, such that the heap condition is satisfied. This is accomplished by recursively shifting the data object of the node $A(i)$ downwards in the subtree, by one level per shift, until the heap condition in the subtree rooted at $A(i)$ is satisfied.

At each step, the most significant among the node $A(i)$ and its children (possibly none, one, or two) is determined. If $A(i)$ is the most significant one (e.g. when $A(i)$ is a leaf), then the subtree rooted at $A(i)$ satisfies the heap condition. In this case, the routine `heapify` terminates. Otherwise, one of its children, the node $A(i_{\max})$, $i_{\max} \in \{2i, 2i+1\}$, is the most significant one. In this case, the data object in $A(i)$ is swapped with the one in $A(i_{\max})$, before `heapify` is recursively applied on the subtree rooted at $A(i_{\max})$. The running time of `heapify` on a subtree of size n is $\mathcal{O}(1)$, for fixing the largest node at each step plus one possible swap, times the number of steps. But the number of steps is bounded above by the height $\log_2(n)$ of the subtree rooted at $A(i)$. Therefore, `heapify` requires $\mathcal{O}(\log(n))$ operations in the worst case.

Algorithm 5 (Heapify).

function `heapify`(A, i);
$\ell = 2i$; $r = 2i + 1$;
if $(\ell \leq n)$ *and* $(A(\ell) > A(i))$
  then $i_{\max} = \ell$;
  else $i_{\max} = i$;
if $(r \leq n)$ *and* $(A(r) > A(i_{\max}))$
  then $i_{\max} = r$;
if $(i_{\max} \neq i)$
  then *swap* $A(i)$ *and* $A(i_{\max})$;
    `heapify`$(A, i_{\max})$;

2.5.2 Building a Heap

The function `heapify` can be used for building a heap from any given array A of size n. In order to see this, first note that the nodes $A(\lfloor n/2 \rfloor + 1), \ldots, A(n)$ are the leaves of the tree. The (empty) subtrees of these nodes do trivially satisfy the heap condition. Now, `heapify` can be applied bottom-up on the remaining entries $A(i)$, $i = \lfloor n/2 \rfloor, \ldots, 1$, in order to convert the array A into one, whose corresponding binary tree satisfies the heap condition. The resulting function `build-heap` is shown in Algorithm 6.

Algorithm 6 (Build Heap).

```
function build-heap(A);
for i = ⌊n/2⌋,...,1
    heapify(A,i);
```

Now let us discuss the correctness of the routine `build-heap`. This is done by induction on i. Observe that prior to the first iteration $i = \lfloor n/2 \rfloor$ the subtrees rooted at the leaves $A(i)$, $i = \lfloor n/2 \rfloor + 1, \ldots, n$, satisfy the heap condition. At each step i, the heap condition in the subtree rooted at $A(i)$ is maintained: Prior to step i, the subtrees rooted at the (possibly none, one, or two) children of the node $A(i)$ satisfy the heap condition by the induction assumption. After having called the function `heapify` in the i-th iteration, the subtree rooted at $A(i)$ also satisfies the heap condition. Therefore, after the final iteration, $i = 1$, the tree rooted at $A(1)$ satisfies the heap condition.

The computational costs required for building a heap of size n by using `build-heap` can be bounded above by the following simple calculation. This is the number of $\mathcal{O}(n)$ iterations in `build-heap` times the running time of `heapify`. But the running time of `heapify` can uniformly be bounded above by $\mathcal{O}(\log(n))$ operations. Altogether, the running time of `build-heap` can thus be bounded above by $\mathcal{O}(n \log(n))$. We remark, however, that the tighter bound of $\mathcal{O}(n)$ can be proven for the computational costs of `build-heap`, when tighter bounds on the running time of `heapify` are used, see e.g. [33], Section 6.3.

2.5.3 Heapsort

Although we are not mainly interested in sorting data objects, it is useful to briefly discuss `heapsort`, Algorithm 7. On given array $A = A(1, \ldots, n)$ of length n, `heapsort` starts by applying the function `build-heap` on A. Due to the heap condition, the most significant node in A is the root $A(1)$. Now the idea of `heapsort` is quite simple: It first exchanges the data object in the root $A(1)$ with the one in the node $A(n)$. Then, the function `heapify` is applied on the root of the subarray $A^{(1)} = A(1, \ldots, n-1)$, of length $n - 1$, so that the reduced array $A^{(1)}$ satisfies the heap condition. This operation is called *pop*. After $n-1$ pops, the subarray $A^{(n-1)} = A(1)$ of length 1 contains the least significant node of the original array A. Moreover, the data objects in A are sorted in increasing order: $\sigma(A(1)) \leq \sigma(A(2)) \leq \ldots \leq \sigma(A(n))$.

Algorithm 7 (Heapsort).

```
function heapsort(A);
build-heap(A);
for i = n,...,2
    swap A(1) and A(i);
    heapify(A(1,...,i − 1),1);
```

It is easy to see that heapsort has running time $\mathcal{O}(n\log(n))$ in the worst case. Indeed, the initial call to build-heap costs at most $\mathcal{O}(n\log(n))$ time, and the subsequent $n-1$ pops cost at most $\mathcal{O}(\log(n))$ time each. Altogether, heapsort performs, in comparison with other sorting algorithms, remarkably well, especially in view of its running time in the worst case scenario. For further details, we refer to the textbook [33].

2.5.4 Implementing a Priority Queue

A *priority queue* is a data structure for maintaining a set X of data objects according to their significance values. The set X is subject to modifications by the insertion and removal of single data objects, and thus the significances of the data objects are to be updated at run time. Since we are mainly interested in point sets, we let in the following discussion $X = \{x_1, \ldots, x_n\} \subset \mathbb{R}^d$ denote a point cloud of size n. Moreover, each point $x \in X$ bears a significance value $\sigma(x) \in \mathbb{R}$. Now a priority queue needs to be able to return the most significant point $x^* \in X$, satisfying $\sigma(x^*) = \max_{x \in X} \sigma(x)$, at any time. To this end, the efficient implementation of a priority queue requires the following basic operations.

- max(X).
  Returns the most significant point in X.
- remove(X,i).
  Removes the point $x_i \in X$ from X, i.e., X is updated by $X = X \setminus x_i$.
- insert(X,x).
  Inserts a new point x into the set X, i.e., X is updated by $X = X \cup x$.
- update(X,i).
  Updates the significance value of the point $x_i \in X$.

The data structure heap is of enormous utility when it comes to provide the above functionality. First of all, if we store the points in the (dynamic) array $X(1,\ldots,n)$ satisfying the heap condition, the routine max, Algorithm 8, is straightforward and simple to implement at computational costs of merely $\mathcal{O}(1)$ operations.

Algorithm 8 (Max).

```
function max(X);
    return X(1);
```

Note that each of the remaining three routines, `remove`, `insert`, and `update`, modify the point set $X = \{x_1, \ldots, x_n\}$, and thus the corresponding array $X = X(1, \ldots, n)$. Therefore, maintaining a priority queue by using the data structure heap boils down to maintaining the heap condition. In other words, the three routines `remove`, `insert`, and `update` need to be implemented such that they maintain the heap condition in the array X. But this can be accomplished by mainly using the basic routine `heapify`, Algorithm 5. The implementation of the three routines `update`, `insert`, and `remove` are subject of the following discussion.

The function `update`, Algorithm 9, assumes that the significance value $\sigma(x_i)$ in the i-th position $X(i)$ of the input array $X = (1, \ldots, n)$ has just been changed. Under the assumption that X had satisfied the heap condition before the update of $\sigma(x_i)$, the routine `update` serves to modify the array X such that it satisfies the heap condition on output.

Algorithm 9 (Update).

```
function update(X, i);
    heapify(X, i);
    while (i > 1) and (X(i) > X(⌊i/2⌋))
        i = ⌊i/2⌋;
        heapify(X, i).
```

As to the correctness of the function `update`, first note that by the modification of $\sigma(x_i)$ the subtree rooted at $X(i)$ may violate the heap condition. Indeed, this may happen when the significance value $\sigma(x_i)$ is reduced. Due to our assumption on X, the subtrees of the (possible) children of the node $X(i)$ satisfy the heap condition. So by the initial call to `heapify`(X, i), the subtree rooted at $X(i)$ satisfies the heap condition. In case the significance value $\sigma(x_i)$ is increased, the subtree of X rooted at the parent $X(\lfloor i/2 \rfloor)$ may violate the heap condition. In this case, the point in the parent node $X(\lfloor i/2 \rfloor)$ needs to be shifted downwards in the tree by *one* level, before the modified node $X(\lfloor i/2 \rfloor)$ must be updated accordingly, and so on. This iterative process is accomplished by calling `heapify` in the while loop of the routine `update`. At each call of `heapify` therein, the current parent may be shifted downwards by at most *one* level, at $\mathcal{O}(1)$ costs. Therefore, the running time of the while loop in `update` can be bounded above by $\mathcal{O}(\log(n))$, and so can the initial call to `heapify` be bounded above by $\mathcal{O}(\log(n))$. Altogether, the function `update` costs at most $\mathcal{O}(\log(n))$ time.

Algorithm 10 (Remove).

function `remove`(X, i);
  $X(i) = X(n)$;
  $X = X(1, \ldots, n-1)$;
  `update`(X, i).

We remark that the operation `remove`$(X, 1)$ is the aforementioned *pop* of heapsort. It is easy to see that the output $X = X(1, \ldots, n-1)$ of `remove` satisfies the heap condition, provided that the input $X = X(1, \ldots, n)$ satisfies the heap condition. Moreover, the computational costs of `remove` can be bounded above by $\mathcal{O}(\log(n))$ operations.

Algorithm 11 (Insert).

function `insert`(X, x);
  $X(n+1) = x$;
  `update`$(X, n+1)$.

The insertion of a new point $x \in X$ into X by using the routine `insert`, Algorithm 11, leads us to an extension of the array $X = X(1, \ldots, n)$ by letting $X(n+1) = x$ for new entry $X(n+1)$ at the $(n+1)$-st position in X. But the extended array $X = X(1, \ldots, n+1)$ may now violate the heap condition. Indeed, the significance value $\sigma(x)$ of the new point x in $X(n+1)$ may be larger than the significance of its parent $X(\lfloor (n+1)/2 \rfloor)$. The subsequent call to `update`$(X, n+1)$ serves to modify the array X, such that X satisfies the heap condition on output. The running time of the routine `insert` is obviously $\mathcal{O}(\log(n))$ in the worst case.

2.6 Quadtrees

A *quadtree* is a data structure, dating back to Finkel and Bentley [71], which is suitable for computing decompositions of planar domains by recursive subdivision. Moreover, a quadtree serves to efficiently solve proximity problems for scattered data points in the plane. In forthcoming applications, we employ quadtrees (and their generalization) mainly for such purposes. But before we proceed, let us first give a formal definition for the term quadtree.

Definition 6. *A* `quadtree` *is a tree such that each of its nodes has either no children or four children.*

The aim of this section is to explain the utility of the hierarchical data structure quadtree for the purposes of our applications, especially for multilevel algorithms and adaptive domain decompositions, to be discussed in Chapter 5. In the following discussion, we let $\Omega \subset \mathbb{R}^2$ denote a rectangular bounded domain, comprising a finite point cloud $X \subset \Omega$. When building the quadtree, we initially associate the root of the quadtree with these two data objects, Ω and X, before each of those objects is recursively split into subdomains and subsets.

The decomposition of Ω is accomplished by recursively splitting Ω into smaller rectangular subdomains called *cells*. Each cell corresponds to a node in the quadtree. But only the leaves of the quadtree, the *leaf cells*, may be split. This is done as follows. When splitting a leaf cell ω, *four* new subcells $\omega_{00}, \omega_{10}, \omega_{01}, \omega_{11}$ are created, such that their interior are pairwise disjoint and their union is ω. By this operation, the cell ω becomes the parent node of the four new leaves $\omega_{00}, \omega_{10}, \omega_{01}, \omega_{11}$, and these four children are *attached* to its parent ω in the quadtree. In particular, the cell ω is no longer a leaf. Figure 2.7 (left) shows the subdivision of a leaf cell ω into four subcells $\omega_{00}, \omega_{10}, \omega_{01}, \omega_{11}$ of equal area. The update of the subtree, rooted at ω, is also shown in Figure 2.7 (right).

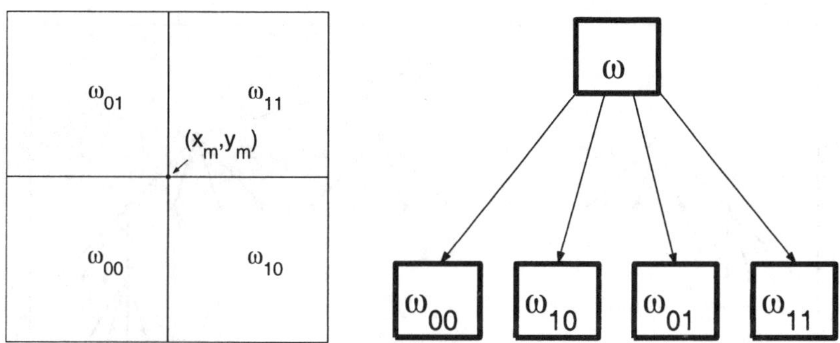

Fig. 2.7. Splitting a cell ω into four subcells $\omega_{00}, \omega_{10}, \omega_{01}, \omega_{11}$.

Likewise, the point cloud X is split by each subdivision of a cell. Any cell ω in the quadtree, initially $\omega = \Omega$, contains a specific portion of X, namely the points of the intersection $X_\omega = X \cap \omega$. Hence, when splitting a cell ω, then the points in X_ω are reattached to the four children $\omega_{00}, \omega_{10}, \omega_{01}, \omega_{11}$, so that

$$X_\omega = X_{\omega_{00}} \cup X_{\omega_{10}} \cup X_{\omega_{01}} \cup X_{\omega_{11}}.$$

Therefore, the whole point set X is the union

$$X = \bigcup_{\omega \in \mathcal{L}} X_\omega$$

of the point sets $X_\omega \subset \omega$, each of which being associated with a leaf ω in the quadtree. We denote the set of the current leaves by $\mathcal{L}$.

Figure 2.8 (top left) shows a rectangular domain $\Omega \subset \mathbb{R}^2$ containing a point set X of size $|X| = 518$. The domain Ω is decomposed into $|\mathcal{L}| = 28$

leaves. Likewise, the point set X is partitioned into the 28 subsets $\{X_\omega\}_{\omega \in \mathcal{L}}$. The structure of the quadtree is also shown in Figure 2.8 (bottom), along with the decomposition of Ω and the partitioning of X (top right).

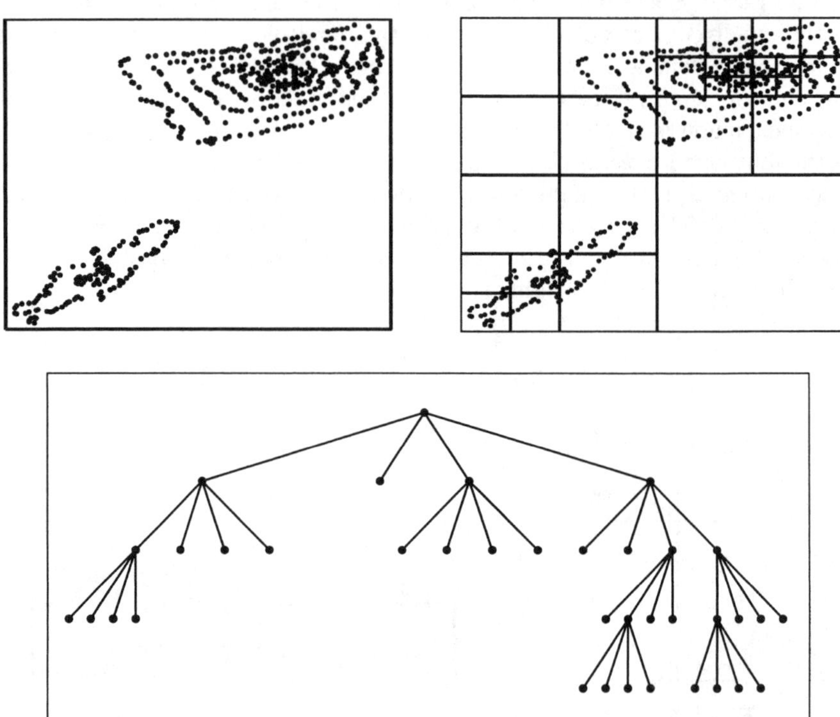

Fig. 2.8. A point set $X \subset \Omega$ comprising 518 points (top left); the subdivision of Ω and partitioning of X (top right); the structure of the corresponding quadtree (bottom).

The splitting of any single leaf ω of the quadtree is done by using the routine `split-cell`, Algorithm 12. The function `split-cell` returns, on given leaf ω, a list $\{\omega_{00}, \omega_{10}, \omega_{01}, \omega_{11}\}$ of new leaves, the four children of ω. Moreover, the points in X_ω are reattached to the cells $\omega_{00}, \omega_{10}, \omega_{01}, \omega_{11}$, before the quadtree data structure is updated by drawing the connectivities between ω and its new children in the graph of the quadtree.

Algorithm 12 (Split Cell).

function `split-cell`(ω);
create four cells $\omega_{00}, \omega_{10}, \omega_{01}, \omega_{11}$;
for $\tilde{\omega} \in \{\omega_{00}, \omega_{10}, \omega_{01}, \omega_{11}\}$
  let $X_{\tilde{\omega}} = X_\omega \cap \tilde{\omega}$;
  attach $\tilde{\omega}$ *to* ω;
return $\{\omega_{00}, \omega_{10}, \omega_{01}, \omega_{11}\}$;

The recursive construction of the entire quadtree is done by the following routine `build-quadtree`, Algorithm 13. The call to `build-quadtree`(Ω) serves to build the whole quadtree.

Algorithm 13 (Build Quadtree).

function `build-quadtree`(ω);
if (ω *is splittable*)
  $\{\omega_{00}, \omega_{10}, \omega_{01}, \omega_{11}\}$ =`split-cell`(ω);
  for $\tilde{\omega} \in \{\omega_{00}, \omega_{10}, \omega_{01}, \omega_{11}\}$
    `build-quadtree`($\tilde{\omega}$);

In each recursion of `build-quadtree`, it is first checked whether the input cell ω is *splittable*. But this requires a splitting criterion for cells. In our applications, we employ different splitting criteria. In order to make a concrete example, one criterion depends merely on the size $|X_\omega|$ of the point set X_ω. In this case, a cell ω is referred to as splittable, iff $|X_\omega| \geq n$ for a predetermined number $n \ll |X|$. So when using this particular splitting criterion, the quadtree is built such that none of its leaves contains more than n points from X. For instance, in the situation of Figure 2.8 we chose the value $n = 50$.

We defer the discussion on other splitting criteria to later in this work. But let us finally remark that by the selection of any specific splitting criterion, the domain is essentially split adaptively. Therefore, unlike binary trees of heaps, the resulting quadtree does not necessarily need to be balanced. But in practice, the height h of the quadtree is a priori bounded above, and satisfying $h \sim \log(N)$, with N being the number of points in X.

2.6.1 Generalizations to Higher Dimensions

In this short subsection, the generalization of quadtrees for higher space dimensions d is explained. Moreover, the function `split-cell` of Algorithm 12 and `build-quadtree` of Algorithm 13 are extended. For the special case of three dimensions, $d = 3$, the data structure is called *octtree*. In this case, any node of the octtree has either no children or $2^3 = 8$ children. Moreover, the underlying domain $\Omega \subset \mathbb{R}^3$, a cuboid in $\mathbb{R}^3$, is recursively split into smaller cuboids, eight at each split. The point set $X \subset \Omega$ is split accordingly.

Now let us turn to the case of a general dimension d.

Definition 7. *For any* $d \geq 2$, *a* `generalized quadtree` *is a tree such that each of its nodes has either no children or* 2^d *children.*

In this general setting, a leaf ω, corresponding to a hypercuboid in $\mathbb{R}^d$, is split into 2^d leaves $\omega_{\mathbf{i}}$, $\mathbf{i} = (i_1, \ldots, i_d) \in \{0,1\}^d$, of pairwise disjoint interior and satisfying

$$\omega = \bigcup_{\mathbf{i} \in \{0,1\}^d} \omega_{\mathbf{i}}.$$

But this leads us immediately to the following extensions of the two routines `split-cell` and `build-quadtree`.

Algorithm 14 (Split Cell).

function `split-cell`(ω);
create 2^d *cells* $\omega_{\mathbf{i}}$, $\mathbf{i} \in \{0,1\}^d$;
for $\tilde{\omega} \in \{\omega_{\mathbf{i}} : \mathbf{i} \in \{0,1\}^d\}$
    let $X_{\tilde{\omega}} = X_\omega \cap \tilde{\omega}$;
    attach $\tilde{\omega}$ *to* ω;
return $\{\omega_{\mathbf{i}} : \mathbf{i} \in \{0,1\}^d\}$;

Algorithm 15 (Build Quadtree).

function `build-quadtree`(ω);
if (ω *is splittable*)
    $\{\omega_{\mathbf{i}} : \mathbf{i} \in \{0,1\}^d\}$ =`split-cell`(ω);
    for $\tilde{\omega} \in \{\omega_{\mathbf{i}} : \mathbf{i} \in \{0,1\}^d\}$
        `build-quadtree`($\tilde{\omega}$);

3 Radial Basis Functions

Radial basis functions are traditional and powerful tools for multivariate scattered data interpolation. Much of the material presented in this chapter is essentially needed in the subsequent developments of this work, such as for the multilevel approximation schemes in Chapter 5, and the meshfree simulation of transport processes in Chapter 6.

In the following Section 3.1, it is first explained *how* radial functions are to be used for scattered data interpolation, and *which* such functions are potentially available for this purpose. The latter leads us to a short discussion on *conditionally positive definite* (radial) functions, in Section 3.2.

The utility of radial basis function interpolation is further supported by the discussion in Sections 3.3 and 3.4, where two important optimality properties of the interpolation scheme are explained, optimal recovery and pointwise optimality.

Section 3.5 is devoted to error estimates, and Section 3.6 is concerning the numerical stability of radial basis function interpolation. A short discussion on one critical aspect of radial basis function interpolation, referred to as the *uncertainty relation*, follows in Section 3.7. Loosely speaking, the uncertainty relation says that none of the commonly used radial basis functions manages to combine good approximation behaviour with a numerically stable interpolation process. This dilemma, discovered by Schaback [146], motivates our recent investigations concerning optimal point sampling, which is the subject of our discussion in Section 3.9.

One particular class of radial basis functions, the *polyharmonic splines* of Duchon [47], are treated separately in Section 3.8, where our recent results concerning the approximation order and the numerical stability of local Lagrange interpolation by polyharmonic splines are reviewed.

The final Section 3.10 of this chapter explains *least squares approximation*, which is in many applications an appropriate alternative to plain interpolation, especially in situations where the given data is contaminated with noise.

For a more comprehensive treatment of radial basis functions, we refer to the surveys [22, 50, 51, 103, 133, 147, 148], the monograph [23], and the recent tutorial [97], where also supporting computer exercises are provided. Further supplementary material and MATLAB software can be downloaded from the web site `www.ma.tum.de/primus2001/radial/`.

3.1 Interpolation

In order to explain multivariate scattered data interpolation by radial basis functions, suppose a data vector $f|_X = (f(x_1), \ldots, f(x_N))^T \in \mathbb{R}^N$ of function values, sampled from an unknown function $f : \mathbb{R}^d \to \mathbb{R}$ at a *scattered* finite point set $X = \{x_1, \ldots, x_N\} \subset \mathbb{R}^d$, $d \geq 1$, is given. Scattered data interpolation requires computing a *suitable* interpolant $s : \mathbb{R}^d \to \mathbb{R}$ satisfying $s|_X = f|_X$, i.e.,

$$s(x_j) = f(x_j), \qquad \text{for all } 1 \leq j \leq N. \tag{3.1}$$

To this end, the radial basis function interpolation scheme works with a fixed *radial* function $\phi : [0, \infty) \to \mathbb{R}$, and the interpolant s in (3.1) is assumed to have the form

$$s(x) = \sum_{j=1}^{N} c_j \phi(\|x - x_j\|) + p(x), \qquad p \in \mathcal{P}_m^d, \tag{3.2}$$

where $\|\cdot\|$ is the Euclidean norm on $\mathbb{R}^d$. Moreover, $\mathcal{P}_m^d$ denotes the linear space containing all real-valued polynomials in d variables of degree at most $m-1$, where $m \equiv m(\phi)$ is said to be the *order* of the basis function ϕ. We come back to the dependence between m and ϕ later in Section 3.2. But let us first give some examples for ϕ.

Classical choices for radial basis functions ϕ, along with their order m, are shown in Table 3.1, where for any real argument $x \in \mathbb{R}$, the symbol $\lceil x \rceil$ denotes as usual the smallest integer greater than or equal to x. Moreover, $\lfloor x \rfloor$ denotes the largest integer less than or equal to x.

Among the most popular radial basis functions are the *polyharmonic splines*, which are discussed more detailed in Section 3.8. This class of radial basis functions includes the *thin plate splines*, where $\phi(r) = r^2 \log(r)$ and $m = 2$, which are particularly suited for interpolation from planar scattered data. Further commonly used radial basis functions are given by the *Gaussians*, $\phi(r) = \exp(-r^2)$, the *multiquadrics*, $\phi(r) = (1 + r^2)^{1/2}$ of order $m = 1$, and the *inverse multiquadrics*, $\phi(r) = (1 + r^2)^{-1/2}$, where $m = 0$. Table 3.1 gives a more general form for the (inverse) multiquadrics and their corresponding order m.

3.1.1 Compactly Supported Radial Basis Functions

More recent developments [170, 175] have provided a whole family of *compactly supported* radial basis functions. In this case, we have $m = 0$ for their order, and so the polynomial part in (3.2) is omitted. While the radial basis functions in Table 3.1 can be used in arbitrary space dimension d, the selection of one *suitable* compactly supported ϕ depends on d, see Table 3.2. Since

Table 3.1. Radial basis functions

Radial Basis Function	$\phi(r) =$	Parameters	Order m
Polyharmonic Splines	r^ν $r^{2k}\log(r)$	$\nu > 0, \nu \notin 2\mathbb{N}$ $k \in \mathbb{N}$	$\lceil \nu/2 \rceil$ $k+1$
Gaussians	$\exp(-r^2)$		0
Multiquadrics	$(1+r^2)^\nu$	$\nu > 0, \nu \notin \mathbb{N}$	$\lceil \nu \rceil$
Inverse Multiquadrics	$(1+r^2)^\nu$	$\nu < 0$	0

the dimension d is usually known beforehand, this is no severe restriction, as shall be established below.

Let us further discuss some basics about compactly supported radial basis functions. As to Wendland's functions [170], these are of the form

$$\phi_{d,k}(r) = \begin{cases} p_{d,k}, & \text{for } 0 \leq r \leq 1, \\ 0, & \text{for } r > 1, \end{cases} \tag{3.3}$$

where $p_{d,k}$ is a specific univariate polynomial of degree $\lfloor d/2 \rfloor + 3k + 1$, and so the support $\text{supp}(\phi_{d,k})$ of $\phi_{d,k} : [0, \infty) \to \mathbb{R}$ is normalized to the unit interval $[0, 1]$. Moreover, due to Wendland's construction in [170], the basis function $\phi_{d,k}$ has derivatives up to order $2k$, i.e., $\phi_{d,k} \in C^{2k}(\mathbb{R}^d)$. Possible choices for $\phi_{d,k}$ are listed in the following Table 3.2, where the symbol $\doteq$ denotes equality up to a positive factor, and the *truncated power function* $(\cdot)_+ : \mathbb{R} \to [0, \infty)$ is given by $(x)_+ = x$, for $x > 0$, and $(x)_+ = 0$, for $x \leq 0$.

By their construction, Wendland's radial basis functions $\phi_{d,k}$ are *positive definite* on $\mathbb{R}^d$.

Definition 8. *A continuous radial function $\phi : [0, \infty) \to \mathbb{R}$ is said to be* `positive definite` *on $\mathbb{R}^d$, iff for any finite set $X = \{x_1, \ldots, x_N\}$, $X \subset \mathbb{R}^d$, of pairwise distinct points the matrix*

$$A_{\phi,X} = (\phi(\|x_j - x_k\|))_{1 \leq j,k \leq N} \in \mathbb{R}^{N \times N}$$

is positive definite. We let $\mathbf{PD}_d$ *denote the set of positive definite functions.*

Due to the construction in [175], there exists, for any space dimension d, a positive definite and compactly supported $\phi \in \mathbf{PD}_d$ of the form (3.3). Remarkably enough, Wendland showed that any basis function $\phi_{d,k}$, constructed

Table 3.2. Wendland's compactly supported radial basis functions [170]

Dimension d	Radial Basis Function	Smoothness C^{2k}
$d = 1$	$\phi_{1,0} = (1-r)_+$ $\phi_{1,1} \doteq (1-r)_+^3(3r+1)$ $\phi_{1,2} \doteq (1-r)_+^5(8r^2+5r+1)$	C^0 C^2 C^4
$d \leq 3$	$\phi_{3,0} = (1-r)_+^2$ $\phi_{3,1} \doteq (1-r)_+^4(4r+1)$ $\phi_{3,2} \doteq (1-r)_+^6(35r^2+18r+3)$ $\phi_{3,3} \doteq (1-r)_+^8(32r^3+25r^2+8r+1)$	C^0 C^2 C^4 C^6
$d \leq 5$	$\phi_{5,0} = (1-r)_+^3$ $\phi_{5,1} \doteq (1-r)_+^5(5r+1)$ $\phi_{5,2} \doteq (1-r)_+^7(16r^2+7r+1)$	C^0 C^2 C^4

in [170] (such as any in Table 3.2), has minimal degree among all positive definite functions $\phi \in \mathbf{PD}_d \cap C^{2k}(\mathbb{R}^d)$ of the form (3.3). Moreover, by these properties, $\phi_{d,k}$ in (3.3) is unique up to a positive constant.

3.1.2 Well-Posedness of the Interpolation Problem

Now let us turn to the well-posedness of the interpolation problem (3.1). We distinguish the case, where $m = 0$ from the one where $m > 0$.

First suppose $m = 0$ for the order of the basis function ϕ, such as for the *Gaussians*, the *inverse multiquadrics* (in Table 3.1) and *Wendland's functions* (in Table 3.2). In this case, the interpolant s in (3.2) has the form

$$s(x) = \sum_{j=1}^{N} c_j \phi(\|x - x_j\|). \tag{3.4}$$

By requiring the N interpolation conditions in (3.1), the computation of the unknown coefficients $c = (c_1, \ldots, c_N)^T \in \mathbb{R}^N$ of s in (3.4) amounts to solving the linear equation system

$$A_{\phi,X} \cdot c = f\big|_X. \tag{3.5}$$

Recall that according to Definition 8, the matrix $A_{\phi,X}$ in (3.5) is guaranteed to be positive definite, provided that $\phi \in \mathbf{PD}_d$. In this case, the system (3.5) has a unique solution. This in turn implies the well-posedness of the given interpolation problem already.

Theorem 1. *For $\phi \in \mathbf{PD}_d$, the interpolation problem (3.1) has a unique solution s of the form (3.4).* □

Now let us turn to the case, where $m > 0$ for the order of ϕ. In this case, the interpolant s in (3.2) contains a nontrivial polynomial part, yielding q additional degrees of freedom, where $q = \binom{m-1+d}{d}$ is the dimension of the polynomial space $\mathcal{P}_m^d$. These additional degrees of freedom are usually eliminated by requiring the q vanishing *moment conditions*

$$\sum_{j=1}^{N} c_j p(x_j) = 0, \qquad \text{for all } p \in \mathcal{P}_m^d. \tag{3.6}$$

Altogether, this amounts to solving the linear system

$$\begin{bmatrix} A_{\phi,X} & P_X \\ P_X & 0 \end{bmatrix} \cdot \begin{bmatrix} c \\ d \end{bmatrix} = \begin{bmatrix} f|_X \\ 0 \end{bmatrix}, \tag{3.7}$$

where we let

$$P_X = ((x_j)^\alpha)_{1 \le j \le N; |\alpha| < m} \in \mathbb{R}^{N \times q}, \tag{3.8}$$

and $d = (d_\alpha)_{|\alpha|<m} \in \mathbb{R}^q$ for the coefficients of the polynomial part in (3.2). For any point $x = (x_1, \ldots, x_d)^T \in \mathbb{R}^d$ and multi-index $\alpha = (\alpha_1, \ldots, \alpha_d) \in \mathbb{N}_0^d$, we let $x^\alpha = x_1^{\alpha_1} \cdot \cdots \cdot x_d^{\alpha_d}$ and $|\alpha| = \alpha_1 + \ldots + \alpha_d$.

In order to analyze the existence and uniqueness of a solution of (3.7), we first consider its corresponding *homogeneous* system

$$A_{\phi,X} \cdot c + P_X \cdot d = 0, \tag{3.9}$$

$$P_X^T \cdot c = 0, \tag{3.10}$$

here split into its interpolation conditions (3.9) and moment conditions (3.10). If we multiply the equation (3.9) from left with c^T, and by using the moment conditions (3.10), we immediately obtain the identity

$$c^T \cdot A_{\phi,X} \cdot c = 0. \tag{3.11}$$

Now in order to guarantee the existence of a solution to (3.9),(3.10), we require that the matrix $A_{\phi,X}$ is, for *any* finite set $X = \{x_1, \ldots, x_N\}$ of interpolation points, *positive definite* on the linear subspace of $\mathbb{R}^d$ containing all vectors $c \in \mathbb{R}^N$ satisfying (3.10). This can be restated as

$$c^T \cdot A_{\phi,X} \cdot c > 0, \quad \text{for all } X \text{ and } c \in \mathbb{R}^N \setminus \{0\} \text{ with } P_X^T \cdot c = 0. \tag{3.12}$$

Note that the side condition in (3.12) depends, due to the definition of P_X in (3.8), on the *order* m of the polynomial space $\mathcal{P}_m^d$. For $m = 0$, the side condition in (3.12) is empty, in which case ϕ is, according to Definition 8, positive definite. This altogether leads us to the following definition.

Definition 9. *A continuous radial function* $\phi : [0, \infty) \to \mathbb{R}$ *is said to be* `conditionally positive definite` *of order* m *on* $\mathbb{R}^d$, $\phi \in \mathbf{CPD}_d(m)$, *iff (3.12) holds for all possible choices of finite point sets* $X \subset \mathbb{R}^d$.

As shall be established in the following Section 3.2, for any radial basis function ϕ in Table 3.1, we either have $\phi \in \mathbf{CPD}_d(m)$ or $-\phi \in \mathbf{CPD}_d(m)$, with the corresponding order m given in the last column of Table 3.1. Note that for any pair $m_1, m_2 \in \mathbb{N}_0$ with $m_1 \leq m_2$, we have the inclusion $\mathbf{CPD}_d(m_1) \subset \mathbf{CPD}_d(m_2)$. For any conditionally positive definite ϕ, we say that the *minimal* $m \equiv m(\phi) \in \mathbb{N}_0$ satisfying (3.12) is the *order* of ϕ. Note that $\mathbf{PD}_d = \mathbf{CPD}(0)$, and so the order of any positive definite $\phi \in \mathbf{PD}_d$, is zero.

Now let us return to the above discussion concerning the solvability of the linear system (3.9),(3.10). For $\phi \in \mathbf{CPD}_d(m)$ (or $-\phi \in \mathbf{CPD}_d(m)$), we conclude $c = 0$ directly from (3.11), and so (3.9) becomes $P_X \cdot d = 0$. Therefore, in order to guarantee a unique solution of (3.9),(3.10), it remains to require the *injectivity* of the matrix P_X. But this property depends on the geometry of the interpolation points in X. Indeed, note that the matrix P_X is injective, iff for $p \in \mathcal{P}_m^d$ the implication

$$p(x_j) = 0 \quad \text{for } 1 \leq j \leq N \qquad \Longrightarrow \qquad p \equiv 0 \tag{3.13}$$

holds. In this case, any polynomial in $\mathcal{P}_m^d$ can uniquely be reconstructed from its function values sampled at the points in X. The point set X is then said to be $\mathcal{P}_m^d$*-unisolvent*. Note that the requirement (3.13) for the points in X is rather weak for small m. Indeed, when $m = 0$, the condition is empty, for $m = 1$ it is trivial, and for $m = 2$ the points in X must not lie on a straight line.

We summarize the above discussion as follows.

Theorem 2. *For* $\phi \in \mathbf{CPD}_d(m)$, *the interpolation problem (3.1) has under constraints (3.6) a unique solution* s *of the form (3.2), provided that the interpolation points in* X *are* $\mathcal{P}_m^d$*-unisolvent by satisfying (3.13).* □

Note that radial basis function interpolation is *meshfree*. This key property of radial basis functions is in contrast to many other methods for scattered data interpolation, such as splines over triangulations. Therefore, radial basis function interpolation does not require additional data structures and algorithms for *grid generation*.

In fact, the implementation of the radial basis function interpolation scheme is, for *well-distributed* data sets X of moderate size N, usually a straightforward task, which requires, merely a few standard methods from numerical linear algebra. For extremely large and unevenly distributed sets X, however, a careful preprocessing of the data points X is required. In this case, multilevel approximation schemes are appropriate tools. Various effective multilevel approximation methods are constructed in Chapter 5, where the preprocessing works (partly) with thinning algorithms, to be introduced in the following Chapter 4.

3.2 Conditionally Positive Definite Functions

By the discussion in the previous section, radial basis function interpolation essentially relies on the conditional positive definiteness the chosen basis function ϕ. Indeed, this is one of the key properties of the interpolation scheme. In this subsection, we discuss two alternative ways for the construction and characterization of conditionally positive definite functions.

One technique, dating back to Micchelli [121], works with *completely monotone functions*. The other alternative relies on *generalized Fourier transforms* [91]. We do not intend to discuss these two different techniques in all details. Instead of this, we briefly review relevant results. For a more comprehensive discussion concerning conditionally positive definite functions, we refer to the recent survey [153].

Completely Monotone Functions.

Definition 10. *A function* $\psi \in C^\infty(0,\infty)$ *is called* `completely monotone` *on* $(0,\infty)$, *iff*

$$(-1)^\ell \psi^\ell(r) \geq 0, \qquad \ell = 0, 1, 2, \ldots,$$

holds for all $r \in (0,\infty)$.

Micchelli provides in [121] a sufficient criterion for $\phi \in \mathbf{CPD}_d(m)$, which generalizes an earlier result by Schoenberg [155, 156] for positive definite radial functions. Micchelli also conjectured the necessity of this criterion. This was finally shown by Guo, Hu and Sun in [82]. We summarize the relevant results from [82, 121, 155, 156] by

Theorem 3. *Let* $\phi : [0,\infty) \to \mathbb{R}$ *be a continuous radial function. Moreover, let* $\phi_{\sqrt{}} \equiv \phi(\sqrt{\cdot})$. *Suppose* $\phi_m \equiv (-1)^m \phi_{\sqrt{}}^{(m)}$ *is well-defined and* ϕ_m *is not constant. Then, the following two statements are equivalent.*

(a) $\phi \in \mathbf{CPD}_d(m)$ *for all* $d \geq 1$;
(b) ϕ_m *is completely monotone on* $(0,\infty)$. □

Now, by using Theorem 3, it is easy to show for any ϕ in Table 3.1 that either ϕ or $-\phi$ is conditionally positive definite of order m, with m given in the last column of Table 3.1. Note, however, that the characterization in Theorem 3 applies to *radial* functions only. Moreover, it excludes the construction of *compactly supported* radial basis functions. The latter is due to the Bernstein-Widder theorem [13] (see also [172]) which says that any function $\psi : [0,\infty) \to \mathbb{R}$ is completely monotone on $(0,\infty)$, if and only if it has a Laplace-Stieltjes-type representation of the form

$$\psi(r) = \int_0^\infty \exp(-rs)\, d\mu(s),$$

where μ is monotonically increasing with $\int_0^\infty d\mu(s) < \infty$. Hence, in this case ψ has no zero, and so any $\psi = \phi_m$ in **(b)** of Theorem 3 cannot be compactly supported.

Generalized Fourier Transforms. A different technique for the characterization and construction of (not necessarily radial) functions $\phi \in \mathbf{CPD}_d(m)$, including compactly supported ones, is using (generalized) Fourier transforms, see the recent survey [153, Section 4] (which basically relies on the results in [91]). We do not explain generalized Fourier transforms here, but rather refer to the textbooks [78, 79], where a comprehensive treatment of the relevant technical background is provided.

For the purposes in this subsection, it is sufficient to say that any radial basis function ϕ in Table 3.1 has a *radial* (generalized) Fourier transform $\hat{\phi} \in C(0, \infty)$ satisfying the following two properties.

- $\hat{\phi}(\|\cdot\|)$ is L_1-integrable around infinity, i.e.,

$$\int_{\mathbb{R}^d \setminus B_1(0)} \left|\hat{\phi}(\|\omega\|)\right| d\omega < \infty, \tag{3.14}$$

- $\hat{\phi}(\|\cdot\|)$ has at most an algebraic singularity of *order* $s_0 \in \mathbb{N}_0$ at the origin, such that

$$\int_{B_1(0)} \|\omega\|^{s_0} \hat{\phi}(\|\omega\|)\, d\omega < \infty, \tag{3.15}$$

  holds, with $s_0 \in \mathbb{N}_0$ being minimal in (3.15).

Table 3.3 shows the (generalized) Fourier transforms of the radial basis functions in Table 3.1, along with their order s_0, where $\doteq$ means equality up to a constant factor, and where K_δ denotes the modified Bessel function, see [1] for properties of K_δ. Note that all functions $\hat{\phi}$ in Table 3.1 are *positive* on $(0, \infty)$.

We remark that if ϕ has a Fourier transform $\hat{\phi} \in L_1(\mathbb{R}^d)$ in the classical sense, satisfying

$$\hat{\phi}(\|\omega\|) = \int_{\mathbb{R}^d} \phi(\|x\|) \exp(-ix^T\omega)\, dx,$$

then this *classical* Fourier transform $\hat{\phi}$ coincides with the generalized Fourier transform of ϕ. Examples are given by the Gaussians, the inverse multiquadrics, and Wendland's compactly supported radial basis functions. In this case, we have $s_0 = 0$ for the order of $\hat{\phi}$.

Now let us turn straight to the characterization of conditionally positive definite functions by generalized Fourier transforms. This particular characterization relies on the identity

$$\sum_{j,k=1}^{N} c_j c_k \phi(\|x_j - x_k\|) = (2\pi)^{-d} \int_{\mathbb{R}^d} \hat{\phi}(\|\omega\|) \left| \sum_{j=1}^{N} c_j \exp(-ix_j^T\omega) \right|^2 d\omega, \tag{3.16}$$

which can be established [91] for any $\hat{\phi}$ satisfying (3.14) and (3.15), provided that the *symbol function*

Table 3.3. Generalized Fourier transforms of radial basis functions

Radial Basis Function	$\phi(r) =$	$\hat{\phi}(s) \doteq$	**Order** s_0
Polyharmonic Splines	r^ν $r^{2k}\log(r)$	$s^{-d-\nu}$ s^{-d-2k}	$\lfloor \nu \rfloor + 1$ $2k+1$
Gaussians	$\exp(-r^2)$	$\exp(-s^2/4)$	0
Multiquadrics	$(1+r^2)^\nu$	$K_{d/2+\nu}(s) \cdot s^{-(d/2+\nu)}$	$\lfloor 2\nu \rfloor + 1$
Inverse Multiquadrics	$(1+r^2)^\nu$	$K_{d/2+\nu}(s) \cdot s^{-(d/2+\nu)}$	0

$$\sigma_{c,X}(\omega) = \sum_{j=1}^{N} c_j \exp(-ix_j^T \omega) \tag{3.17}$$

has a zero at the origin of order at least $m = \lceil s_0/2 \rceil$. Note that the latter can be guaranteed by requiring the moment conditions (3.6) with $m = \lceil s_0/2 \rceil$.

Theorem 4. *A continuous radial function $\phi : [0,\infty) \to \mathbb{R}$ is conditionally positive definite on $\mathbb{R}^d$, if ϕ has a continuous non-negative generalized Fourier transform $\hat{\phi} \not\equiv 0$ satisfying (3.14) and (3.15). In this case, $m = \lceil s_0/2 \rceil$ is the order of $\phi \in$ **CPD**$_d(m)$.*

Proof. Let $\hat{\phi}$ satisfy (3.14) and (3.15), and suppose (3.6) with $m = \lceil s_0/2 \rceil$, so that the identity (3.16) holds. By the non-negativity of $\hat{\phi}$, the quadratic form

$$c^T \cdot A_{\phi,X} \cdot c = \sum_{j,k=1}^{N} c_j c_k \phi(\|x_j - x_k\|)$$

appearing in the left hand side of (3.16), is non-negative. Hence it remains to show that $c^T \cdot A_{\phi,X} \cdot c$ vanishes, if and only if $c = 0$. In order to see this, suppose that $c^T \cdot A_{\phi,X} \cdot c$, and thus the right hand side in (3.16), vanishes. In this case, the symbol function $\sigma_{c,X}$ in (3.17) must vanish on an open subset of $\mathbb{R}^d$ with nonempty interior. But then, due to the analyticity of $\sigma_{c,X}$, this implies that the symbol function vanishes identically on $\mathbb{R}^d$, i.e., $\sigma_{c,X} \equiv 0$. Since the points in X are pairwise distinct, and so the exponentials $\exp(-ix_j^T\omega)$ are linearly independent, the latter is true, if and only if $c = 0$. □

3.3 Optimal Recovery

In this section, the suitability of radial basis functions for *optimal recovery* is shown. To this end, we follow along the lines of the variational theory by Madych and Nelson [114, 115, 116]. Due to Madych and Nelson, every *conditionally positive definite* function $\phi \in \mathbf{CPD}_d(m)$ is associated with a *native function space* $\mathcal{F}_\phi$, being equipped with a semi-norm $|\cdot|_\phi$, such that the corresponding interpolation scheme (as discussed in Section 3.1) is *optimal* in $\mathcal{F}_\phi$. More precisely, for any $f \in \mathcal{F}_\phi$ and $X = \{x_1, \dots, x_N\}$, the (unique) interpolant $s_{f,X}$ of the form (3.2), with $s_{f,X}\big|_X = f\big|_X$, lies in the native space $\mathcal{F}_\phi$, and it satisfies

$$|s_{f,X}|_\phi \le |f|_\phi, \qquad \text{for } f \in \mathcal{F}_\phi. \tag{3.18}$$

In other words, the interpolant $s_{f,X}$ minimizes the energy $|\cdot|_\phi$ among all interpolants $g \in \mathcal{F}_\phi$ satisfying $g\big|_X = f\big|_X$. This property of the interpolation scheme is referred to as *optimal recovery* in the sense of Micchelli, Rivlin, and Winograd [122].

We remark that the results in [114, 115, 116] are not necessarily restricted to *radial* functions $\phi \in \mathbf{CPD}_d(m)$. In fact, the following construction of $\mathcal{F}_\phi$ merely relies on the conditional positive definiteness of ϕ. Nevertheless, in order to avoid unnecessary detours, we keep on using the assumption that $\phi \in \mathbf{CPD}_d(m)$ is radial.

In order to see that ϕ provides an optimal interpolation scheme, let us first rewrite the form of the interpolant s in (3.2). Recall that for any linear combination

$$\lambda = \sum_{x \in X} \lambda_x \delta_x \tag{3.19}$$

of translates of the Dirac δ-functional, the expression

$$\lambda * \phi = \sum_{x \in X} \lambda_x \phi(\|\cdot - x\|)$$

is the *convolution product* between λ and ϕ. Therefore, any s in (3.2) has the form

$$s_{\lambda,p} = \lambda * \phi + p, \qquad \lambda \in \mathcal{L}_m^\perp, p \in \mathcal{P}_m^d, \tag{3.20}$$

where

$$\mathcal{L}_m^\perp = \left\{ \lambda = \sum_{x \in X} \lambda_x \delta_x \; : \; X \subset \mathbb{R}^d, |X| < \infty, \lambda\big|_{\mathcal{P}_m^d} = 0 \right\}$$

contains all *finite* linear combinations of the form (3.19) which are vanishing on $\mathcal{P}_m^d$. Now all possible interpolants of the above form (3.20) constitute a linear space

$$\mathcal{R}_\phi = \left\{ s_{\lambda,p} \; : \; \lambda \in \mathcal{L}_m^\perp, p \in \mathcal{P}_m^d \right\},$$

which is referred to as the *recovery space* associated with ϕ.

Note that the dual space $\mathcal{L}_m^\perp$ is an inner product space with

$$(\lambda, \mu)_\phi = \lambda(\mu * \phi), \qquad \text{for all } \lambda, \mu \in \mathcal{L}_m^\perp. \tag{3.21}$$

That this is an inner product is due to the identity

$$\lambda(\mu * \phi) = \lambda\left(\mu^y \phi(\cdot - y)\right) = \lambda^x \mu^y \phi(x - y)$$

and the conditional positive definiteness of ϕ. As an immediate consequence, $\mathcal{R}_\phi$ is equipped with the semi-inner product

$$(s_{\lambda,p}, s_{\mu,q})_\phi = (\lambda, \mu)_\phi, \qquad \text{for all } \lambda, \mu \in \mathcal{L}_m^\perp, p, q \in \mathcal{P}_m^d. \tag{3.22}$$

This leads us to the *Madych-Nelson space*

$$\mathcal{F}_\phi = \left\{\lambda * \phi : \lambda \in \overline{\mathcal{L}_m^\perp}\right\} \bigoplus \mathcal{P}_m^d, \tag{3.23}$$

where $\overline{\mathcal{L}_m^\perp}$ denotes the topological closure of $\mathcal{L}_m^\perp$.

In the remainder of this section, it is shown that $\mathcal{F}_\phi$ constitutes an *optimal recovery space* for the interpolation process explained in Section 3.1, i.e., (3.18) holds for any function $f \in \mathcal{F}_\phi$. The starting point for doing so is the following result proven in the seminal paper [115] by Madych and Nelson.

Theorem 5. *Let $\phi \in \mathbf{CPD}_d(m)$. Then, $\mathcal{F}_\phi$ satisfies the following three conditions.*

(a) *the null space of the semi-norm $|\cdot|_\phi$ is $\mathcal{P}_m^d$;*
(b) *the factor space $\mathcal{F}_\phi / \mathcal{P}_m^d$ is a Hilbert space;*
(c) *$(\lambda * \phi, f)_\phi = \lambda(f)$ for all $f \in \mathcal{F}_\phi$ and $\lambda \in \mathcal{L}_m^\perp$.*

As shown in [115], the above properties **(a)**-**(c)** uniquely determine the function space $\mathcal{F}_\phi$ among all subspaces of $C(\mathbb{R}^d, \mathbb{R})$.

Note that the above property **(c)** holds for all $f \in \mathcal{R}_\phi$ and $\lambda \in \mathcal{L}_m^\perp$. Indeed, by combining (3.22) with (3.21), we obtain for any $f = s_{\mu,q} \in \mathcal{R}_\phi$ the identity

$$(\lambda * \phi, s_{\mu,q})_\phi = (\lambda, \mu)_\phi = \lambda(\mu * \phi) = \lambda(s_{\mu,q}) \qquad \text{for all } \lambda \in \mathcal{L}_m^\perp.$$

By continuity, we obtain **(c)** for all $f \in \mathcal{F}_\phi$. As to the proofs of **(a)** and **(b)**, we refer the reader to [115].

Now, the validity of property **(c)** in Theorem 5 leads us to the desired optimality (in the sense of (3.18)) of the radial basis function recovery scheme.

Theorem 6. *Let $\phi \in \mathbf{CPD}_d(m)$. Then, the Pythagoras theorem*

$$|f|_\phi^2 = |s_f|_\phi^2 + |f - s_f|_\phi^2 \tag{3.24}$$

holds for all $f \in \mathcal{F}_\phi$.

Proof. Note that for $s_f = \lambda * \phi + p \in \mathcal{R}_\phi$,

$$(s_f, g)_\phi = 0 \quad \text{for all } g \text{ with } \lambda(g) = 0 \tag{3.25}$$

holds by statement **(c)** of Theorem 5. But this implies

$$(s_f, f - s_f)_\phi = 0, \tag{3.26}$$

i.e., s_f is the orthogonal projection of $f \in \mathcal{F}_\phi$ onto $\mathcal{R}_\phi$. This immediately implies (3.24) which completes our proof. □

We finally remark that, due to [115], an alternative representation for $\mathcal{F}_\phi$ is given by

$$\mathcal{F}_\phi = \left\{ f \in C(\mathbb{R}^d) \,:\, |\lambda(f)| \leq C_f \|\lambda\|_\phi \text{ for all } \lambda \in \overline{\mathcal{L}_m^\perp} \text{ with } C_f \geq 0 \right\}. \tag{3.27}$$

Hence, the Madych-Nelson space $\mathcal{F}_\phi$ is the largest linear space on which functionals from $\overline{\mathcal{L}_m^\perp}$ are continuous, see [93] for details. For a recent account on native spaces of radial functions, we refer to the papers [149, 150].

3.4 Pointwise Optimality

In this section, another optimality property of the radial basis function interpolation scheme is discussed, cf. [176].

3.4.1 Lagrange Representation of the Interpolant

In the following discussion of this section, it is convenient to work with the *Lagrange representation*

$$s_{f,X}(x) = \sum_{j=1}^{N} \lambda_j(x) f(x_j) \tag{3.28}$$

of the interpolant $s \equiv s_{f,X}$ in (3.2), where the *Lagrange basis functions* $\lambda_1(x), \ldots, \lambda_N(x)$ satisfy

$$\lambda_j(x_k) = \begin{cases} 1, & \text{for } j = k, \\ 0, & \text{for } j \neq k \end{cases} \qquad 1 \leq j, k \leq N, \tag{3.29}$$

and so $s\big|_X = f\big|_X$.

For a fixed point $x \in \mathbb{R}^d$, the vectors

$$\lambda(x) = (\lambda_1(x), \ldots, \lambda_N(x))^T \in \mathbb{R}^N \quad \text{and} \quad \mu(x) = (\mu_1(x), \ldots, \mu_q(x))^T \in \mathbb{R}^q$$

are the unique solution of the linear system

$$\begin{bmatrix} A_{\phi,X} & P_X \\ P_X^T & 0 \end{bmatrix} \cdot \begin{bmatrix} \lambda(x) \\ \mu(x) \end{bmatrix} = \begin{bmatrix} R_{\phi,X}(x) \\ S(x) \end{bmatrix}, \tag{3.30}$$

where $R_{\phi,X}(x) = (\phi(\|x - x_j\|))_{1 \le j \le N} \in \mathbb{R}^N$ and $S(x) = (x^\alpha)_{|\alpha|<m} \in \mathbb{R}^q$. We abbreviate the linear system (3.30) as

$$\mathbf{A} \cdot \nu(x) = \mathbf{b}(x)$$

by letting

$$\mathbf{A} = \begin{bmatrix} A_{\phi,X} & P_X \\ P_X^T & 0 \end{bmatrix}, \quad \nu(x) = \begin{bmatrix} \lambda(x) \\ \mu(x) \end{bmatrix}, \quad \mathbf{b}(x) = \begin{bmatrix} R_{\phi,X}(x) \\ S(x) \end{bmatrix}.$$

This allows us to combine the two alternative representations for s in (3.28) and (3.2) by

$$\begin{aligned} s(x) &= < \lambda(x), f|_X > \\ &= < \nu(x), f_X > \\ &= < \mathbf{A}^{-1} \cdot \mathbf{b}(x), f_X > \\ &= < \mathbf{b}(x), \mathbf{A}^{-1} \cdot f_X > \\ &= < \mathbf{b}(x), b >, \end{aligned} \tag{3.31}$$

where $< \cdot, \cdot >$ denotes the inner product of the Euclidean space $\mathbb{R}^d$, and where we let

$$f_X = \begin{bmatrix} f|_X \\ 0 \end{bmatrix} \in \mathbb{R}^{N+q} \quad \text{and} \quad b = \begin{bmatrix} c \\ d \end{bmatrix} \in \mathbb{R}^{N+q}$$

for the right hand side and the solution of the linear system (3.7).

3.4.2 Pointwise Error Bounds

In this subsection, we derive for a fixed $x \in \mathbb{R}^d$ bounds for the pointwise error

$$\epsilon_x(f) = |f(x) - s_{f,X}(x)|. \tag{3.32}$$

Note that $\epsilon_x(f) = 0$ for all $x \in \mathbb{R}^d$, provided that $f \in \mathcal{P}_m^d$. Indeed, since the interpolation scheme reconstructs polynomials from $\mathcal{P}_m^d$, we have

$$p(x) = \sum_{j=1}^{N} \lambda_j(x) p(x_j), \qquad \text{for all } p \in \mathcal{P}_m^d. \tag{3.33}$$

Thus,

$$\epsilon_x = \delta_x - \sum_{j=1}^{N} \lambda_j(x)\delta_{x_j} \in \mathcal{L}_m^{\perp}.$$

By using part **(c)** of Theorem 5, (3.32) can be rewritten as

$$|\epsilon_x(f)| = |(\epsilon_x * \phi, f)_\phi|$$

and therefore by using the Cauchy-Schwarz inequality we obtain the pointwise error bound

$$|f(x) - s_{f,X}(x)| \leq |f|_\phi \cdot \|\epsilon_x\|_\phi. \tag{3.34}$$

Note that this bound is sharp, with equality in (3.34) being attained by the function

$$f = \epsilon_x * \phi = \phi(\|\cdot - x\|) - \sum_{j=1}^{N} \lambda_j(x)\phi(\|\cdot - x_j\|).$$

Now, the norm of the error functional ϵ_x can be expressed as

$$\|\epsilon_x\|_\phi^2 = \phi(0) - 2\sum_{j=1}^{N} \lambda_j(x)\phi(\|x - x_j\|) + \sum_{j,k=1}^{N} \lambda_j(x)\lambda_k(x)\phi(\|x_j - x_k\|). \tag{3.35}$$

The resulting function $\pi_{\phi,X}(x) \equiv \|\epsilon_x\|_\phi$ is referred to as the *power function* of the interpolation scheme. The following theorem gives us four useful alternative representations for the power function.

Theorem 7. *The power function $\pi_{\phi,X} = \|\epsilon_x\|_\phi$ can be expressed as follows.*

(a) $\pi_{\phi,X}^2(x) = \phi(0) - 2\lambda^T(x)R_{\phi,X}(x) + \lambda^T(x) \cdot A_{\phi,X} \cdot \lambda(x)$;

(b) $\pi_{\phi,X}^2(x) = \overline{\lambda}(x)^T \cdot A_{\phi,x\cup X} \cdot \overline{\lambda}(x)$, *where* $\overline{\lambda}^T(x) = (-1, \lambda_1(x), \ldots, \lambda_N(x))$;

(c) $\pi_{\phi,X}^2(x) = \phi(0) - \lambda^T(x)R_{\phi,X}(x) - \mu^T(x)S(x)$;

(d)

$$\pi_{\phi,X}^2(x) = \phi(0) - \begin{bmatrix} \lambda(x) \\ \mu(x) \end{bmatrix}^T \cdot \begin{bmatrix} A_{\phi,X} & P_X \\ P_X^T & 0 \end{bmatrix} \cdot \begin{bmatrix} \lambda(x) \\ \mu(x) \end{bmatrix}.$$

Proof. Note that the representation in **(a)** follows directly from (3.35). Moreover, **(a)** immediately implies **(b)** by using the identity

$$\overline{\lambda}(x)^T \cdot A_{\phi,x\cup X} \cdot \overline{\lambda}(x) = \left[-1, \lambda^T(x)\right] \cdot \begin{bmatrix} \phi(0) & R_{\phi,X}^T(x) \\ R_{\phi,X}(x) & A_{\phi,X} \end{bmatrix} \cdot \begin{bmatrix} -1 \\ \lambda(x) \end{bmatrix}$$

$$= \phi(0) - 2\lambda^T(x)R_{\phi,X}(x) + \lambda^T(x)A_{\phi,X}\lambda(x).$$

As to **(c)**, expanding (3.30) gives us

$$A_{\phi,X}\lambda(x) + P_X\mu(x) = R_{\phi,X}(x)$$
$$P_X^T\lambda(x) = S(x)$$

and therefore

$$\lambda^T(x) \cdot A_{\phi,X} \cdot \lambda(x) = \lambda^T(x)R_{\phi,X}(x) - \mu^T(x)S(x),$$

which, by using the representation **(a)**, implies **(c)**. Finally, rewriting **(c)** as

$$\pi_{\phi,X}^2(x) = \phi(0) - \begin{bmatrix} \lambda(x) \\ \mu(x) \end{bmatrix}^T \cdot \begin{bmatrix} R_{\phi,X}(x) \\ S(x) \end{bmatrix}$$

gives us, by using (3.30), the representation **(d)**. □

Corollary 1. *For $\phi \in \mathbf{PD}_d$, the power function $\pi_{\phi,X}$ can be bounded by*

$$0 \le \pi_{\phi,X}^2(x) \le \phi(0). \tag{3.36}$$

Proof. The representation **(c)** in Theorem 7 yields

$$\pi_{\phi,X}^2(x) = \phi(0) - \lambda^T(x)A_{\phi,X}\lambda(x)$$

for the special case $\phi \in \mathbf{PD}_d$. But in this case, $m = 0$, the matrix $A_{\phi,X}$ is positive definite and this gives us the upper bound on $\pi_{\phi,X}^2$ in (3.36). Since $\pi_{\phi,X}^2$ is non-negative, this completes our proof. □

Next, we consider minimizing the error norm $\pi_{\phi,X} \equiv \pi_{\phi,\lambda,X}$ by variation of the coefficients $\lambda \equiv \lambda(x) \in \mathbb{R}^N$ in (3.28), and under constraints (3.33). This leads us to the quadratic optimization problem

$$\min_{\lambda\in\mathbb{R}^N} \pi_{\phi,\lambda,X}^2 = \min_{\lambda\in\mathbb{R}^N} \left(\phi(0) - 2\lambda^T R_{\phi,X}(x) + \lambda^T \cdot A_{\phi,X} \cdot \lambda\right)$$

with linear side conditions

$$P_X^T\lambda = S(x).$$

A solution $\lambda^* \in \mathbb{R}^N$ of this minimization problem satisfies

$$A_{\phi,X}\lambda^* + P_X\mu^* = R_{\phi,X}(x)$$
$$P_X^T\lambda^* = S(x),$$

with the Lagrange-multipliers $\mu^* \in \mathbb{R}^q$. Note that this is the linear system (3.30) whose unique solution $\lambda^* \equiv \lambda(x) = (\lambda_1(x), \ldots, \lambda_N(x)) \in \mathbb{R}^N$ is given by the cardinal functions satisfying (3.29). Altogether, we obtain

Theorem 8. *Let $X = \{x_1, \ldots, x_N\} \subset \mathbb{R}^d$ and $\phi \in \mathbf{CPD}_d(m)$. Then, the interpolant $s_{f,X}$ in (3.28) satisfying (3.33) is, for any fixed $x \in \mathbb{R}^d$, the unique minimizer of the pointwise error bound (3.34) among all quasi-interpolants of the form*

$$s(x) = \sum_{j=1}^{N} \tilde{\lambda}_j(x) f(x_j),$$

satisfying

$$p(x) = \sum_{j=1}^{N} \tilde{\lambda}_j(x) p(x_j), \qquad \text{for all } p \in \mathcal{P}_m^d. \qquad \square$$

3.5 Error Estimates

In the following discussion, available bounds on the error $\|f - s_{f,X}\|_{L_\infty(\Omega)}$ are provided, where $\Omega \subset \mathbb{R}^d$ is a bounded and open domain comprising X, i.e., $X \subset \Omega$. Moreover, it is assumed that $\Omega \subset \mathbb{R}^d$ satisfies an *interior cone condition.*

Definition 11. *The domain $\Omega \subset \mathbb{R}^d$ is said to satisfy an* `interior cone condition`*, iff there exists an angle θ and a radius $\varrho > 0$, such that for every $x \in \Omega$ there exists a unit vector $\xi \equiv \xi(x)$ satisfying*

$$C(x, \xi, \theta, \varrho) = \left\{x + ty \,:\, y \in \mathbb{R}^d, \|y\| = 1, y^T\xi \geq \cos(\theta), t \in [0, \varrho]\right\} \subset \Omega,$$

i.e., the cone $C(x, \xi, \theta, \varrho)$ is entirely contained in Ω.

Let us first discuss pointwise error estimates. Available bounds on the error (3.32), for $x \in \Omega$, are proven in [115, 116, 176]. These pointwise error estimates rely on upper bounds on the power function's value $\pi_{\phi,X}(x)$. Such bounds depend on the *local fill distance*

$$h_{\varrho,X}(x) = \max_{y \in B_\varrho(x)} d_X(y)$$

of X around x, where $B_\varrho(x) = \{y : \|y - x\| \leq \varrho\}$, denotes the closed ball around x of radius ϱ. Moreover, recall from (2.4), that

$$d_X(y) = \min_{x \in X} \|y - x\| \tag{3.37}$$

is the Euclidean distance between the point y and the point set X.

Now, for all commonly used radial basis functions, particularly for those listed in Table 3.1, available pointwise error estimates have the form

$$\pi^2_{\phi,X}(x) \leq C \cdot F_\phi(h_{\varrho,X}(x)), \tag{3.38}$$

where $F_\phi : [0,\infty) \to [0,\infty)$ is a monotonically increasing function with $F_\phi(0) = 0$, depending merely on ϕ. For any radial basis function ϕ in Table 3.1, its corresponding F_ϕ (see also [148]) is listed in Table 3.4, where the symbol $\doteq$ stands for equality up to a positive constant.

Note that (3.38) yields in combination with (3.34) the pointwise error bound

$$|f(x) - s_{f,X}(x)| \le C \cdot |f|_\phi \cdot F_\phi^{1/2}(h_{\varrho,X}(x)).$$

It can be shown that these pointwise error bounds carry over to uniform bounds in the domain Ω, which leads us to error estimates depending on the *fill distance*

$$h_{X,\Omega} = \max_{y \in \Omega} d_X(y) \tag{3.39}$$

of X in Ω, i.e.,

$$\|f - s_{f,X}\|_{L_\infty(\Omega)} \le C \cdot |f|_\phi \cdot F_\phi^{1/2}(h_{X,\Omega}) \tag{3.40}$$

for every $f \in \mathcal{F}_\phi$. For further details, we refer to [148, 152].

3.6 Numerical Stability

Now let us turn to the *numerical stability* of radial basis function interpolation. As explained in [151], the numerical stability of the linear system (3.7) is dominated by the *spectral condition number* of the matrix $A_{\phi,X}$ in (3.7). Since $A_{\phi,X}$ is, due to (3.12), positive definite on the kernel $\ker(P_X^T) \subset \mathbb{R}^N$, there are positive eigenvalues $\sigma_{\max}$ and $\sigma_{\min}$ of $A_{\phi,X}$ satisfying

$$\sigma_{\max}\|c\|^2 \ge c^T \cdot A_{\phi,X} \cdot c \ge \sigma_{\min}\|c\|^2, \qquad \text{for all } c \in \mathbb{R}^N \text{ with } P_X^T c = 0.$$

Now the condition of the linear system (3.7) is given by the ratio $\sigma_{\max}/\sigma_{\min}$. Hence, for the sake of numerical stability, one wants to keep this ratio small. But this requires both upper bounds on $\sigma_{\max}$ and lower bounds on $\sigma_{\min}$. While small upper bounds on $\sigma_{\max}$ are readily available for any $\phi \in \mathbf{CPD}_d(m)$, see [151], it turns out that small values of $\sigma_{\min}$ typically spoil the stability of the interpolation. The latter is also supported by numerical experiments.

Therefore, the discussion in the literature [6, 7, 127, 128, 146, 151] on the numerical stability of radial basis function interpolation is focusing on lower bounds for the smallest eigenvalue $\sigma_{\min}$. The resulting estimates have the form

$$c^T \cdot A_{\phi,X} \cdot c \ge \sigma_{\min}\|c\|^2 \ge G_\phi(q_X)\|c\|^2, \tag{3.41}$$

for all $c \in \ker(P_X^T)$, where

$$q_X = \min_{x \in X} d_{X \setminus x}(x)$$

is the *separation distance* of the point set X. Moreover, $G_\phi : [0,\infty) \to [0,\infty)$ is a monotonically increasing function with $G_\phi(0) = 0$. The form of the corresponding functions G_ϕ, belonging to the radial basis functions in Table 3.1, are listed in Table 3.4, see [148] for more details.

Table 3.4. Radial basis functions: convergence rates and condition numbers

Radial Basis Function	$\phi(r) =$	$F_\phi(h) \doteq$	$G_\phi(q) \doteq$
Polyharmonic Splines	r^ν $r^{2k}\log(r)$	h^ν h^{2k}	q^ν q^{2k}
Gaussians	$\exp(-r^2)$	$\exp(-\alpha/h)$	$\exp(-\beta/q^2)$
(Inverse) Multiquadrics	$(1+r^2)^{\nu/2}$	$\exp(-\alpha/h)$	$q^\nu\exp(-\beta/q)$

In summary, the numerical stability is one critical aspect of radial basis function interpolation. In fact, the numerical stability may become a severe problem in relevant applications, where the distribution of the points in X is very heterogeneous, or where the data set X is extremely large. In situations of large data, multilevel approximation schemes are appropriate tools. In the discussion of Chapter 5, we propose various efficient multilevel schemes, which widely help to avoid such stability problems. As regards the numerical stability in situations where the data set X is rather small, we propose a preconditioning of the system (3.7) for the special case of polyharmonic spline interpolation. This is done in Subsection 3.8.3.

But let us first review a more general discussion concerning radial basis function interpolation, known as the *uncertainty principle*, which explains the conflict between numerical stability and good approximation behaviour.

3.7 Uncertainty Principle

As observed by Schaback [146], there is no commonly used radial basis function which combines good approximation behaviour with a small condition number of the collocation matrix in (3.7). This dilemma is in [146] referred to as the *uncertainty relation* of radial basis function interpolation. This phenomenon has extensively been explained in the survey [148].

In order to combine the results of the previous two Sections 3.5 and 3.6, first note that both the approximation quality and the stability of radial basis function interpolation relies, by the values of q_X and $h_{X,\Omega}$, on the geometry of the interpolation points in X. On the one hand, for the sake of numerical stability, the separation distance q_X should be not too small. This is due to the lower bound

$$\sigma_{\min} \geq G_\phi(q_X) > 0$$

in (3.41) on the smallest eigenvalue $\sigma_{\min}$. On the other hand, for the sake of good approximation quality, it is desired to keep the fill distance $h_{X,\Omega}$ small. This is due to the upper bound in (3.40) on the error $\|f - s_{f,X}\|_{L_\infty(\Omega)}$.

However, it is obviously not possible to minimize $h_{X,\Omega}$ and to maximize q_X at the same time. In fact, due to Theorem 14 in Section 3.9, the relation

$$q_X \leq \sqrt{\frac{2(d+1)}{d}} \cdot h_{X,\Omega} \tag{3.42}$$

holds for any (admissible) point set $X \subset \mathbb{R}^d$ and any space dimension d. This observation already explains why we cannot combine small upper bounds $F_\phi(h_{X,\Omega})$ on $\|f - s_{f,X}\|_{L_\infty(\Omega)}$ in (3.40) with large lower bounds $G_\phi(q_X)$ on $\sigma_{\min}$ in (3.41).

The following arguments, due to [146], serve to bridge the gap between *pointwise* error bounds and bounds on eigenvalues. On the one hand, by part **(b)** of Theorem 7, we obtain for any $x \in \Omega$ the error bound

$$\pi^2_{\phi,X}(x) = \overline{\lambda}(x)^T \cdot A_{\phi,x\cup X} \cdot \overline{\lambda}(x) \leq F_\phi(h_{\varrho,X}(x)).$$

On the other hand, we have

$$\begin{aligned}
\overline{\lambda}(x)^T \cdot A_{\phi,x\cup X} \cdot \overline{\lambda}(x) &\geq G_\phi(q_{x\cup X}) \cdot \|\overline{\lambda}(x)\|^2 \\
&= G_\phi(q_{x\cup X}) \cdot \left(1 + \|\lambda(x)\|^2\right) \\
&\geq G_\phi(q_{x\cup X}).
\end{aligned}$$

Altogether, this implies

$$G_\phi(q_{x\cup X}) \leq F_\phi(h_{\varrho,X}(x)), \qquad \text{for any } x \in \Omega, \tag{3.43}$$

which shows that for small arguments $q_{X\cup x} \approx h \approx h_{\varrho,X}(x)$ in (3.43) one cannot have a small error bound $F_\phi(h)$ without obtaining a small lower bound $G_\phi(h)$ on the smallest eigenvalue.

3.8 Polyharmonic Splines

In this section, details on the *polyharmonic splines*, see Table 3.1, often also referred to as *surface splines*, are explained. The utility of polyharmonic splines for multivariate interpolation was established by Duchon [46, 47, 48]. In order to discuss the particular setting of Duchon, let us be more specific about the choice of the basis function ϕ. According to [46, 47, 48], we assume from now the form

$$\phi_{d,k}(r) = \begin{cases} r^{2k-d}\log(r), & \text{for } d \text{ even}, \\ r^{2k-d}, & \text{for } d \text{ odd}, \end{cases}$$

for the polyharmonic splines, where k is required to satisfy $2k > d$. According to Table 3.1 (last column), the order of $\phi_{d,k}$ is $m = k - \lceil d/2 \rceil + 1$.

Recall that the inclusion $\mathbf{CPD}_d(m_1) \subset \mathbf{CPD}_d(m_2)$, for $m_1 \leq m_2$, allows us to also work with any order greater than m. In order to comply with Duchon's setting, we replace the *minimal* choice $m = k - \lceil d/2 \rceil + 1$ by $k \geq m$. Therefore, we let from now $m = k$ for the order of $\phi_{d,k} \in \mathbf{CPD}_d(m)$. We come back with an explanation concerning this particular choice for m in Subsection 3.8.1.

With using $m = k$, the resulting interpolant in (3.2) has the form

$$s(x) = \sum_{j=1}^{N} c_j \phi_{d,k}(\|x - x_j\|) + \sum_{|\alpha|<k} d_\alpha x^\alpha. \tag{3.44}$$

We remark that the polyharmonic spline $\phi_{d,k}$ is the fundamental solution of the k-th iterated Laplacian, i.e.,

$$\Delta^k \phi_{d,k}(\|x\|) = c\delta_x.$$

For instance, for $d = k = 2$, the thin plate spline $\phi_{2,2}(r) = r^2\log(r)$ solves the *biharmonic equation*

$$\Delta\Delta\phi_{2,2}(\|x\|) = c\delta_x,$$

and in this case, the interpolant s in (3.44) has the form

$$s(x) = \sum_{j=1}^{N} c_j \|x - x_j\|^2 \log(\|x - x_j\|) + d_1 + d_2\xi + d_3\eta, \tag{3.45}$$

where we let ξ and η denote the two coordinates of $x = (\xi, \eta)^T \in \mathbb{R}^2$.

Finally, we remark that for the univariate case, $d = 1$, the polyharmonic spline $\phi_{1,k} = r^{2k-1}$, $k \geq 1$, coincides with the *natural spline* of order $2k$.

3.8.1 Optimal Recovery in Beppo Levi Spaces

Recall the discussion in Section 3.3 concerning optimal recovery of radial basis function interpolation in native function spaces. In this subsection, we introduce *Beppo Levi spaces*, being the optimal recovery spaces of polyharmonic spline interpolation.

Due to fundamental results in the seminal papers [46, 47, 48] of Duchon and [117, 118, 119] of Meinguet, for a fixed finite point set $X \subset \mathbb{R}^d$, an interpolant s in (3.44) minimizes the energy

$$|f|^2_{\mathbf{BL}^k(\mathbb{R}^d)} = \int_{\mathbb{R}^d} \sum_{|\alpha|=k} \binom{k}{\alpha} (D^\alpha f)^2 \, dx, \qquad \binom{k}{\alpha} = \frac{k!}{\alpha_1! \cdot \dots \cdot \alpha_d!}, \tag{3.46}$$

among all functions f of the *Beppo Levi space*

$$\mathbf{BL}^k(\mathbb{R}^d) = \left\{ f \in C(\mathbb{R}^d) \,:\, D^\alpha f \in L^2(\mathbb{R}^d) \text{ for all } |\alpha| = k \right\} \subset C(\mathbb{R}^d)$$

satisfying $f\big|_X = s\big|_X$. So the Beppo Levi space $\mathbf{BL}^k(\mathbb{R}^d)$ is equipped with the semi-norm $|\cdot|_{\mathbf{BL}^k(\mathbb{R}^d)}$, whose kernel is the polynomial space $\mathcal{P}_k^d$. The latter explains why we use order $m = k$ rather than the *minimal* choice $m = k - \lceil d/2 \rceil + 1$. In this case, the Beppo Levi space $\mathbf{BL}^k(\mathbb{R}^d)$ is the *optimal recovery space* $\mathcal{F}_\phi$ for the polyharmonic splines $\phi_{d,k}$. Note that $\mathbf{BL}^k(\mathbb{R}^d)$ is the Sobolev space $\mathcal{H}^k(\mathbb{R}^d)$.

When working with thin plate splines, $\phi_{2,2}(r) = r^2 \log(r)$, in two dimensions we have

$$|f|^2_{\mathbf{BL}^2(\mathbb{R}^2)} = \int_{\mathbb{R}^2} \left(\frac{\partial^2 f}{\partial \xi^2}\right)^2 + 2\left(\frac{\partial^2 f}{\partial \xi \partial \eta}\right)^2 + \left(\frac{\partial^2 f}{\partial \eta^2}\right)^2 d\xi \, d\eta, \qquad \text{for } f \in \mathbf{BL}^2(\mathbb{R}^2).$$

Note that the semi-norm $|\cdot|_{\mathbf{BL}^2(\mathbb{R}^2)}$ is the *bending energy* of a thin plate of infinite extent, and this explains the naming of *thin plate splines.*

3.8.2 Approximation Order

In this subsection, we prove approximation orders for *local* Lagrange interpolation by polyharmonic splines. But let us first recall approximation orders for the *global* case from the discussion in Section 3.5. To this end, we review available bounds on the L_∞-error $\|f - s_{f,X}\|_{L_\infty(\Omega)}$ on a bounded domain Ω. For further technical details on the error analysis, we refer to [152].

Global Approximation Order. Starting point for the discussion in [152] is an *algebraic decay condition*

$$\gamma_1 \|\omega\|^{-d-s_\infty} \le \hat{\phi}(\|\omega\|) \le \gamma_2 \|\omega\|^{-d-s_\infty}, \qquad \text{for } \|\omega\| \to \infty, \tag{3.47}$$

around infinity on the (generalized) Fourier transform $\hat{\phi}$ of $\phi \in \mathbf{CPD}_d(m)$ in $\mathbb{R}^d$, where $0 < \gamma_1 \le \gamma_2$ in (3.47) are suitable constants and $s_\infty > 0$ is the *order* of the decay. For the polyharmonic spline $\phi_{d,k}$, the algebraic decay condition (3.47) is satisfied with $s_\infty = 2k - d$.

Due to [176], the power function $\pi_{\phi,X}$ can, with assuming (3.47), for any $\phi \in \mathbf{CPD}_d(m)$, uniformly be bounded above by

$$\pi_{\phi,X}(x) \leq C \cdot h_{X,\Omega}^{s_\infty/2}, \qquad \text{for all } x \in \Omega,$$

provided that the domain Ω satisfies an interior cone condition. This leads us directly to the desired result concerning the (global) approximation order of scattered data interpolation by polyharmonic splines.

Theorem 9. *Let Ω be a bounded and open domain satisfying an interior cone condition. Then, there exist constants h_0, C, such that for any finite point set $X \subset \Omega$ satisfying $h_{X,\Omega} \leq h_0$ and any function $f \in \mathbf{BL}^k(\mathbb{R}^d)$ the error bound*

$$\|f - s\|_{L_\infty(\Omega)} \leq C \cdot |f|_{\mathbf{BL}^k(\mathbb{R}^d)} h_{X,\Omega}^{k-d/2}$$

holds, where s is the unique polyharmonic spline interpolant in (3.44), using $\phi_{d,k}$, satisfying $s|_X = f|_X$.

Hence, in this sense, the *global* approximation order for interpolation by using the polyharmonic spline $\phi_{d,k}$ is $p = k - d/2$ with respect to the Beppo Levi space $\mathbf{BL}^k(\mathbb{R}^d)$.

Local Approximation Order. In the following discussion of this subsection, we analyze the approximation order of *local* polyharmonic spline interpolation. We remark that this analysis, in combination with the subsequent investigation concerning the stability of local polyharmonic spline interpolation, is of primary interest for applications in multiscale flow simulation, which is the subject of the discussion in Section 6.4 of Chapter 6.

As regards the local approximation order, we consider solving, for some fixed point $x_0 \in \mathbb{R}^d$ and any $h > 0$, the *scaled* interpolation problem

$$f(x_0 + hx_i) = s^h(x_0 + hx_i), \quad 1 \leq i \leq n, \tag{3.48}$$

where $X = \{x_1, \ldots, x_n\} \subset \mathbb{R}^d$ is a $\mathcal{P}_k^d$-unisolvent point set of *moderate* size, i.e., n is small. Moreover, s^h denotes the unique polyharmonic spline interpolant of the form

$$s^h(hx) = \sum_{j=1}^{n} c_j^h \phi_{d,k}(\|hx - hx_j\|) + \sum_{|\alpha|<k} d_\alpha^h (hx)^\alpha \tag{3.49}$$

satisfying (3.48). The discussion in this subsection is dominated by the following definition.

Definition 12. *Let s^h denote the polyharmonic spline interpolant, using $\phi_{d,k}$, satisfying (3.48). We say that the* `approximation order` *of local polyharmonic spline interpolation at $x_0 \in \mathbb{R}^d$ and with respect to the function space $\mathcal{F}$ is p, iff for any $f \in \mathcal{F}$ the asymptotic bound*

$$|f(x_0 + hx) - s^h(x_0 + hx)| = \mathcal{O}(h^p), \quad h \to 0,$$

holds for any $x \in \mathbb{R}^d$, and any finite $\mathcal{P}_k^d$-unisolvent point set $X \subset \mathbb{R}^d$.

For the sake of notational simplicity, we let from now $x_0 = 0$, which is, due to the shift-invariance of the interpolation scheme, without loss of generality.

Note that the coefficients $c^h = (c_1^h, \ldots, c_n^h)^T \in \mathbb{R}^n, d^h = (d_\alpha^h)_{|\alpha|<k} \in \mathbb{R}^q$ of the interpolant s^h in (3.49) are solving the linear system

$$\begin{bmatrix} A_h & P_h \\ P_h^T & 0 \end{bmatrix} \cdot \begin{bmatrix} c^h \\ d^h \end{bmatrix} = \begin{bmatrix} f\big|_{hX} \\ 0 \end{bmatrix}, \tag{3.50}$$

where we let

$$\begin{aligned} A_h &= (\phi_{d,k}(\|hx_i - hx_j\|)_{1\le i,j\le n} \in \mathbb{R}^{n\times n}, \\ P_h &= ((hx_i)^\alpha)_{1\le i\le n; |\alpha|<k} \in \mathbb{R}^{n\times q}, \\ f\big|_{hX} &= (f(hx_i))_{1\le i\le n} \in \mathbb{R}^n. \end{aligned}$$

We abbreviate the above linear system (3.50) as

$$\mathbf{A}_h \cdot b^h = f_h, \tag{3.51}$$

i.e., for notational brevity, we let

$$\mathbf{A}_h = \begin{bmatrix} A_h & P_h \\ P_h^T & 0 \end{bmatrix}, \quad b^h = \begin{bmatrix} c^h \\ d^h \end{bmatrix}, \quad \text{and} \quad f_h = \begin{bmatrix} f\big|_{hX} \\ 0 \end{bmatrix}.$$

Recall from the discussion in Section 3.4 that any interpolant s^h satisfying (3.48) has a Lagrange-type representation of the form

$$s^h(hx) = \sum_{i=1}^n \lambda_i^h(hx) f(hx_i), \tag{3.52}$$

corresponding to the one in (3.28), where

$$\sum_{i=1}^n \lambda_i^h(hx) p(hx_i) = p(hx), \qquad \text{for all } p \in \mathcal{P}_k^d, \tag{3.53}$$

due to the reconstruction of polynomials in $\mathcal{P}_k^d$.

Moreover, for $x \in \mathbb{R}^d$, the vector $\lambda^h(hx) = (\lambda_1^h(hx), \ldots, \lambda_n^h(hx))^T \in \mathbb{R}^n$ is, together with $\mu^h(hx) = (\mu_\alpha^h(hx))_{|\alpha|<k} \in \mathbb{R}^q$, the unique solution of the linear system

$$\begin{bmatrix} A_h & P_h \\ P_h^T & 0 \end{bmatrix} \cdot \begin{bmatrix} \lambda^h(hx) \\ \mu^h(hx) \end{bmatrix} = \begin{bmatrix} R_h(hx) \\ S_h(hx) \end{bmatrix}, \tag{3.54}$$

where

$$R_h(hx) = (\phi_{d,k}(\|hx - hx_j\|))_{1\le j\le n} \in \mathbb{R}^n$$

$$S_h(hx) = ((hx)^\alpha)_{|\alpha|<k} \in \mathbb{R}^q .$$

It is convenient to abbreviate the system (3.54) as

$$\mathbf{A}_h \cdot \nu^h(hx) = \mathbf{b}_h(hx),$$

i.e., we let

$$\mathbf{A}_h = \begin{bmatrix} A_h & P_h \\ P_h^T & 0 \end{bmatrix}, \quad \nu^h(hx) = \begin{bmatrix} \lambda^h(hx) \\ \mu^h(hx) \end{bmatrix}, \quad \mathbf{b}_h(hx) = \begin{bmatrix} R_h(hx) \\ S_h(hx) \end{bmatrix}.$$

Starting with the Lagrange representation of s^h in (3.52), we obtain

$$\begin{aligned} s^h(hx) &= < \lambda^h(hx), f\big|_{hX} > \\ &= < \nu^h(hx), f_h > \\ &= < \mathbf{A}_h^{-1} \cdot \mathbf{b}_h(hx), f_h > \qquad (3.55) \\ &= < \mathbf{b}_h(hx), \mathbf{A}_h^{-1} \cdot f_h > \\ &= < \mathbf{b}_h(hx), b_h >, \end{aligned}$$

see the identity (3.31). This in particular combines the two alternative representations for s^h in (3.52) and (3.49).

The following lemma, proven in [102], plays a key role in the following discussion. It states that the Lagrange basis of the polyharmonic spline interpolation scheme is invariant under uniform scalings. As established in the recap of the proof from [102] below, this result mainly relies on the (generalized) homogeneity of $\phi_{d,k}$.

Lemma 2. *For any $h > 0$, let $\lambda^h(hx)$ be the solution in (3.54). Then,*

$$\lambda^h(hx) = \lambda^1(x), \qquad \textit{for every } x \in \mathbb{R}^d.$$

Proof. For fixed $X = \{x_1, \ldots, x_n\} \subset \mathbb{R}^d$, and any $h > 0$, let

$$\mathcal{R}^h_{\phi,X} = \left\{ \sum_{j=1}^n c_j^h \phi_{d,k}(\|\cdot - hx_j\|) + p \,:\, p \in \mathcal{P}_k^d, \sum_{j=1}^n c_j^h q(x_j) = 0 \text{ for all } q \in \mathcal{P}_k^d \right\}$$

denote the *recovery space* of all possible polyharmonic spline interpolants of the form (3.49) satisfying (3.48).

In what follows, we show that $\mathcal{R}_{\phi,X}^h$ is a scaled version of $\mathcal{R}_{\phi,X}^1$, so that

$$\mathcal{R}_{\phi,X}^h = \left\{\sigma_h(s) \,:\, s \in \mathcal{R}_{\phi,X}^1\right\}, \tag{3.56}$$

where the dilatation operator σ_h is given by $\sigma_h(s) = s(\cdot/h)$. This then implies that, due to the unicity of the interpolation in either space, $\mathcal{R}_{\phi,X}^h$ or $\mathcal{R}_{\phi,X}^1$, their Lagrange basis functions must coincide by satisfying $\lambda^h = \sigma_h(\lambda^1)$, as stated above.

In order to show that $\mathcal{R}_{\phi,X}^h = \sigma_h(\mathcal{R}_{\phi,X}^1)$, we distinguish the special case where d is even from the one where d is odd. If the space dimension d is odd, then $\mathcal{R}_{\phi,X}^h = \sigma_h(\mathcal{R}_{\phi,X}^1)$ follows immediately from the homogeneity of $\phi_{d,k}$, where $\phi_{d,k}(hr) = h^{2k-d}\phi_{d,k}(r)$.

Now suppose that d is even. In this case we have

$$\phi_{d,k}(hr) = h^{2k-d}\left(\phi_{d,k}(r) + r^{2k-d}\log(h)\right).$$

Therefore, any function $s^h \in \mathcal{R}_{\phi,X}^h$ has, for some $p \in \mathcal{P}_k^d$, the form

$$s^h(hx) = h^{2k-d}\left(\sum_{j=1}^n c_j^h \phi_{d,k}(\|x - x_j\|) + \log(h)q(x)\right) + p(x),$$

where we let

$$q(x) = \sum_{j=1}^n c_j^h \|x - x_j\|^{2k-d}.$$

In order to see that s^h is contained in $\sigma_h(\mathcal{R}_{\phi,X}^1)$, it remains to show that the degree of the polynomial q is at most $k-1$. To this end, we rewrite q as

$$q(x) = \sum_{j=1}^n c_j^h \sum_{|\alpha|+|\beta|=2k-d} c_{\alpha,\beta}\cdot x^\alpha (x_j)^\beta = \sum_{|\alpha|+|\beta|=2k-d} c_{\alpha,\beta}\cdot x^\alpha \sum_{j=1}^n c_j^h (x_j)^\beta,$$

for some coefficients $c_{\alpha,\beta} \in \mathbb{R}$ with $|\alpha| + |\beta| = 2k - d$. Due to the vanishing moment conditions

$$\sum_{j=1}^n c_j^h p(hx_j) = 0, \qquad \text{for all } p \in \mathcal{P}_k^d,$$

for the coefficients $c_1^h, \ldots, c_n^h$, this implies that the degree of q is at most $2k - d - k = k - d < k$. Therefore, $s^h \in \sigma_h(\mathcal{R}_{\phi,X}^1)$, and so $\mathcal{R}_{\phi,X}^h \subset \sigma_h(\mathcal{R}_{\phi,X}^1)$. The inclusion $\mathcal{R}_{\phi,X}^1 \subset \sigma_h^{-1}(\mathcal{R}_{\phi,X}^h)$ can be proven accordingly.

Altogether, we find that $\mathcal{R}_{\phi,X}^h = \sigma_h(\mathcal{R}_{\phi,X}^1)$ for any d, which completes our proof. □

Now let us draw important conclusions on the approximation order of local polyharmonic spline interpolation with respect to C^k. To this end, regard for $f \in C^k$, any $x \in \mathbb{R}^d$ and $h > 0$, the k-th order Taylor polynomial

$$T^h_{f,hx}(y) = \sum_{|\alpha|<k} \frac{1}{\alpha!} D^\alpha f(hx)(y - hx)^\alpha \tag{3.57}$$

of f around hx. By using

$$f(hx) = T^h_{f,hx}(hx_i) - \sum_{0<|\alpha|<k} \frac{1}{\alpha!} D^\alpha f(hx)(hx_i - hx)^\alpha, \qquad \text{for all } 1 \le i \le n,$$

in combination with (3.52) and (3.53), we obtain the identity

$$f(hx) - s^h(hx) = \sum_{i=1}^n \lambda_i^h(hx) \left[T^h_{f,hx}(hx_i) - f(hx_i)\right].$$

Now due to Lemma 2, the *Lebesgue constant*

$$\Lambda = \sup_{h>0} \sum_{i=1}^n |\lambda_i^h(hx)| = \sum_{i=1}^n |\lambda_i^1(x)|$$

is bounded, locally around the origin $x_0 = 0$, and therefore we can conclude

$$|f(hx) - s^h(hx)| = \mathcal{O}(h^k), \quad h \to 0.$$

Altogether, this yields the following result.

Theorem 10. *The approximation order of local polyharmonic spline interpolation, using $\phi_{d,k}$, with respect to C^k is $p = k$.* □

We remark that the above Theorem 10 generalizes a previous result in [84] concerning the local approximation order of thin plate spline interpolation in the plane.

Corollary 2. *The approximation order of local thin plate spline interpolation, using $\phi_{2,2} = r^2 \log(r)$, with respect to C^2 is $p = 2$.* □

3.8.3 Numerical Stability

This subsection is devoted to the construction of a numerically stable algorithm for the evaluation of polyharmonic spline interpolants. Recall that the stability of an algorithm always depends on the conditioning of the given problem. For a more general discussion on the relevant principles and concepts from error analysis, especially the *condition number* of a given *problem* versus the *stability* of a *numerical algorithm*, we recommend the textbook [87].

In order to briefly explain the conditioning of polyharmonic spline interpolation, let $\Omega \subset \mathbb{R}^d$ denote a compact domain comprising $X = \{x_1, \ldots, x_n\}$, i.e., $X \subset \Omega$, the $\mathcal{P}_k^d$-unisolvent set of interpolation points. Now recall that the condition number of an interpolation operator $\mathcal{I} : C(\Omega) \to C(\Omega)$, $\Omega \subset \mathbb{R}^d$, w.r.t. the L_∞-norm $\|\cdot\|_{L_\infty(\Omega)}$, is the smallest number κ_∞ satisfying

$$\|\mathcal{I}f\|_{L_\infty(\Omega)} \leq \kappa_\infty \cdot \|f\|_{L_\infty(\Omega)} \quad \text{for all } f \in C(\Omega).$$

Thus, κ_∞ is the operator norm of $\mathcal{I}$ w.r.t. the norm $\|\cdot\|_{L_\infty(\Omega)}$. In the situation of polyharmonic spline interpolation, the interpolation operator $\mathcal{I}_{d,k} : C(\Omega) \to C(\Omega)$, returns, for any given argument $f \in C(\Omega)$ the polyharmonic spline interpolant $\mathcal{I}_{d,k}(f) = s_f \in C(\Omega)$ of the form (3.44) satisfying $s_f\big|_X = f\big|_X$. The following result is useful for the subsequent discussion on the stability of local interpolation by polyharmonic splines.

Theorem 11. *The condition number κ_∞ of interpolation by polyharmonic splines is given by the Lebesgue constant*

$$\Lambda(\Omega, X) = \max_{x \in \Omega} \sum_{i=1}^{n} |\lambda_i(x)|. \tag{3.58}$$

Proof. For $f \in C(\Omega)$, let $f\big|_X$ be given, and let $s_f = \mathcal{I}_{d,k}(f) \in C(\Omega)$ denote the interpolant of the form (3.44) satisfying $f\big|_X = s_f\big|_X$. Using the Lagrange-type representation

$$s_f(x) = \sum_{i=1}^{n} \lambda_i(x) f(x_i)$$

of s_f, we obtain

$$\|\mathcal{I}_{d,k}f\|_{L_\infty(\Omega)} = \|s_f\|_{L_\infty(\Omega)} \leq \max_{x \in \Omega} \sum_{i=1}^{n} |\lambda_i(x)| \cdot |f(x_i)| \leq \Lambda(\Omega, X) \cdot \|f\|_{L_\infty(\Omega)}$$

for all $f \in C(\Omega)$, and therefore $\kappa_\infty \leq \Lambda(\Omega, X)$.

In order to see that $\kappa_\infty \geq \Lambda(\Omega, X)$, suppose that the maximum of $\Lambda(\Omega, X)$ in (3.58) is attained at $x^* \in \Omega$. Moreover, let $g \in C(\Omega)$ denote any function satisfying $g(x_i) = \text{sign}(\lambda_i(x^*))$, for all $1 \leq i \leq n$, and $\|g\|_{L_\infty(\Omega)} = 1$. Then, we obtain

$$\|\mathcal{I}_{d,k}g\|_{L_\infty(\Omega)} \geq (\mathcal{I}_{d,k}g)(x^*) = \sum_{i=1}^{n} \lambda_i(x^*) g(x_i) = \sum_{i=1}^{n} |\lambda_i(x^*)| = \Lambda(\Omega, X)$$

and thus $\|\mathcal{I}_{d,k}g\|_{L_\infty(\Omega)} \geq \Lambda(\Omega, X)\|g\|_{L_\infty(\Omega)}$. But this implies $\Lambda(\Omega, X) \leq \kappa_\infty$. Altogether, $\kappa_\infty = \Lambda(\Omega, X)$, which completes our proof. □

Theorem 11, in combination with Lemma 2 of the previous Subsection 3.8.2, immediately yields the following important result concerning the stability of interpolation by polyharmonic splines.

Theorem 12. *The absolute condition number of polyharmonic spline interpolation is invariant under rotations, translations and uniform scalings.*

Proof. Interpolation by polyharmonic splines is invariant under rotations and translations. It is easy to see that this property carries over to the absolute condition number. In order to see that $\kappa_\infty \equiv \kappa_\infty(\Omega, X)$ is also invariant under uniform scalings, let $\Omega^h = \{hx : x \in \Omega\}$ and $X^h = \{hx : x \in X\}$. Then, we obtain

$$\Lambda(\Omega^h, X^h) = \max_{hx \in \Omega^h} \sum_{i=1}^n \lambda_i^h(hx) = \max_{x \in \Omega} \sum_{i=1}^n \lambda_i(x) = \Lambda(\Omega, X)$$

which shows that $\kappa_\infty(\Omega^h, X^h) = \kappa_\infty(\Omega, X)$. □

Now let us turn to the construction of a numerically stable algorithm for evaluating the polyharmonic spline interpolant s^h satisfying (3.48). To this end, we require that the given interpolation problem (3.48) is well-conditioned. Note that according to Theorem 12, this requirement depends on the geometry of the interpolation points X w.r.t. the center x_0, but not on the scale h.

However, the spectral condition number of the matrix $\mathbf{A}_h$ in (3.51) depends on h. The following rescaling can be viewed as a simple way of preconditioning the matrix $\mathbf{A}_h$ for very small h. In order to evaluate the polyharmonic spline interpolant s^h satisfying (3.48), we prefer to work with the representation

$$s^h(hx) = < \mathbf{b}_1(x), \mathbf{A}_1^{-1} \cdot f_h >, \tag{3.59}$$

which immediately follows from the identity (3.55) and the scale-invariance of the Lagrange basis, Lemma 2. Due to (3.59) we can evaluate s^h at hx by solving the linear system

$$\mathbf{A}_1 \cdot \mathbf{b} = f_h. \tag{3.60}$$

The solution $\mathbf{b} \in \mathbb{R}^{n+q}$ in (3.60) then yields the coefficients of $s^h(hx)$ w.r.t. the basis functions in $\mathbf{b}_1(x)$.

By working with the representation (3.59) for s^h instead of the one in (3.49), we can avoid solving the linear system (3.51). This is useful insofar as the linear system (3.51) is ill-conditioned for very small h, but well-conditioned for sufficiently large h. The latter relies on earlier results due to Narcowich and Ward [128], where it is shown that the spectral norm of the matrix A_h^{-1} is bounded above by a monotonically decreasing function of the separation distance q_{hX} of the point set $hX = \{hx_1, \ldots, hx_n\}$. This in turns implies that one should, for the sake of numerical stability, avoid solving the system (3.51) directly for very small h, see Section 3.6. For further details on this, we refer to [128] and the more general discussion provided by the recent paper [151] of Schaback.

3.9 Optimal Point Sampling

This section concerns the construction and characterization of point sets X, whose *uniformity* $\rho_{X,\Omega} = q_X/h_{X,\Omega}$ is, for a compact domain $\Omega \subset \mathbb{R}^d$, maximal. Our motivation for the following discussion is mainly given by the uncertainty relation of Section 3.7. In fact, the results in Section 3.7 suggest to consider the variation of the points in X for the purpose of improving the performance of radial basis function interpolation in terms of numerical stability and approximation quality. This requires balancing the two quantities q_X and $h_{X,\Omega}$, such that q_X is large and $h_{X,\Omega}$ is small, see the discussion around (3.42). This explains why we want to maximize the uniformity $\rho_{X,\Omega} = q_X/h_{X,\Omega}$. Note that neither $h_{X,\Omega}$ nor q_X depends on the selected basis function ϕ. Nonetheless their values contribute, for all commonly used radial basis functions, significantly to the method's performance [95].

The following theorem provides one useful isoperimetric property of *regular simplices.* Recall that a simplex is said to be regular, iff all of its edge lengths are equal. Hence, for the special case $d = 2$, a regular simplex is an equilateral triangle.

Theorem 13. *Let T denote a nondegenerate d-dimensional simplex, V_T the set of its $d+1$ vertices, and r_T the radius of its circumsphere S_T. Furthermore let*

$$J_d = \sqrt{\frac{d}{2(d+1)}}.$$

Then we have

$$q_{V_T} \leq r_T/J_d$$

where equality holds if and only if T is a regular simplex. □

The above theorem is a well-known result from discrete computational geometry proven by Rankin [138] in the context of *spherical codes* (see also [32, Chapter 1, Section 2.6]). The notation J_d for the reciprocal value of the uniformity $\rho_{V_{\Delta_d},\Delta_d}$ of a regular d-simplex Δ_d is dedicated to H. Jung [105] (see [24, Chapter 2, Section 11]).

In the following discussion we assume for any point set X that the domain Ω contains at least one Voronoi vertex of X, i.e., $\Omega \cap \mathcal{V}_X$ is non-empty, where we let $\mathcal{V}_X$ denote the set of Voronoi points of X (see Section 2.3 of Chapter 2). In this case, the set X is said to be *admissible*, and we collect all such finite point sets in

$$\mathcal{X} = \{X = \{x_1, \ldots, x_N\} \subset \mathbb{R}^d \ : \ \Omega \cap \mathcal{V}_X \text{ not empty } \}.$$

We say that X is *optimally distributed* in Ω, iff X maximizes the uniformity $\rho_{X,\Omega}$ among all point sets $Y \in \mathcal{X}$ by satisfying

$$\rho_{X,\Omega} = \sup_{Y \in \mathcal{X}} \rho_{Y,\Omega}.$$

The following theorem provides an upper bound on the uniformity for point sets in $\mathcal{X}$.

Theorem 14. *Let $X \in \mathcal{X}$. Then, the uniformity of X in Ω can be bounded above by*

$$\rho_{X,\Omega} \leq J_d^{-1}.$$

Proof. Let $\mathcal{D}_X$ be the Delaunay triangulation of the point set X. Then, the shortest distance between two distinct points from X is given by the length of the shortest edge in $\mathcal{D}_X$, so that $q_X = \min_{T \in \mathcal{D}_X} q_{V_T}$ holds. Moreover, we have

$$h_{X,\Omega} = \max_{y \in \Omega} d_X(y) \geq \max_{v \in \mathcal{V}_X \cap \Omega} d_X(v) = \max_{\substack{T \in \mathcal{D}_X \\ c_T \in \Omega}} r_T. \tag{3.61}$$

By using Theorem 13 we find

$$\rho_{X,\Omega} = \frac{q_X}{h_{X,\Omega}} \leq \frac{\min_{T \in \mathcal{D}_X} q_{V_T}}{\max_{\substack{T \in \mathcal{D}_X \\ c_T \in \Omega}} r_T} \leq \max_{\substack{T \in \mathcal{D}_X \\ c_T \in \Omega}} \frac{q_{V_T}}{r_T} \leq J_d^{-1}, \tag{3.62}$$

which completes our proof. □

The construction of optimally distributed point sets relies on the following necessary condition.

Theorem 15. *Let $X \in \mathcal{X}$ be an optimally distributed set in Ω by satisfying $\rho_{X,\Omega} = J_d^{-1}$. Then, every simplex $T \in \mathcal{D}_X$ whose circumsphere's center c_T lies in Ω is regular.*

Proof. From (3.61) we conclude $h_{X,\Omega} \geq r_T$, and therefore, like in (3.62), we obtain

$$\rho_{X,\Omega} \leq q_{V_T}/r_T \leq J_d^{-1}.$$

But then, $\rho_{X,\Omega} = J_d^{-1}$ particularly implies $q_{V_T}/r_T = J_d^{-1}$. According to Theorem 13, the simplex T is then regular. □

The above result suggests using regular simplices for the construction of optimally distributed point sets. We further discuss this by providing the following observation.

Theorem 16. *Let X be a finite point set, such that every simplex of its Delaunay triangulation $\mathcal{D}_X$ is regular. Then, the set X is optimally distributed in its convex hull $[X]$.*

Proof. Without loss of generality let the edge lengths of the simplices be normalized, such that $q_T = 1$ holds for every $T \in \mathcal{D}_X$. Under this assumption and by using

$$h_{V_T,T} = J_d \quad \text{for every } T \in \mathcal{D}_X$$

we find for $\Omega = \text{conv}(X)$ the identity

$$h_{X,\Omega} = \max_{y \in \Omega} d_X(y) = \max_{T \in \mathcal{D}_X} \max_{y \in T} d_X(y) = \max_{T \in \mathcal{D}_X} h_{V_T,T} = J_d$$

which, in combination with Theorem 14, completes our proof. □

For the special case of two dimensions, where $J_2^{-1} = \sqrt{3}$, optimal point sets can be constructed by using the *hexagonal lattice*

$$\mathcal{H}_2 = \{jv_1 + kv_2 \, : \, j,k \in \mathbb{Z}\} \subset \mathbb{R}^2,$$

which is spanned by $v_1 = (1,0)$ and $v_2 = (1/2, \sqrt{3}/2)$.

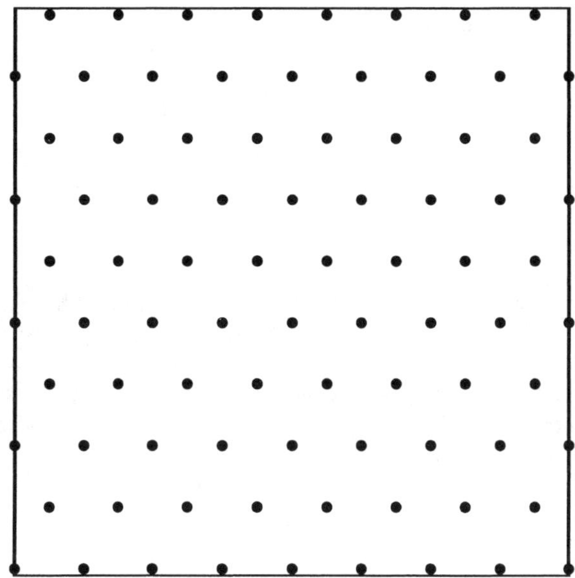

Fig. 3.1. An optimal point set $X \subset [0,1]^2$ satisfying $\rho_{X,\Omega} = \sqrt{3}$.

Figure 3.1 shows one example for a planar point set $X \subset \mathcal{X}$ which is optimal in the unit square $\Omega = [0,1]^2$ by satisfying $\rho_{X,\Omega} = \sqrt{3}$. Further details on the construction of such point sets are explained in [95], where moreover numerical results concerning the performance of radial basis function interpolation are provided.

3.10 Least Squares Approximation

This final section is devoted to *least squares approximation*, an alternative approach for scattered data fitting other than plain interpolation. Least squares approximation makes sense especially in situations where the given data is contaminated with noise.

For the purpose of explaining this particular approximation scheme, let a finite point set $X = \{x_1, \ldots, x_N\} \subset \mathbb{R}^d$ and data values $f\big|_X$ be given. Moreover, suppose $Y = \{y_1, \ldots, y_n\}$ is a subset of X, $Y \subset X$, whose size $|Y| = n$ is much smaller than the size $|X| = N$ of X, i.e., $n \ll N$. As to the construction of a *suitable* subset Y, this can be accomplished by using scattered data filtering, a selection scheme which is explained in Section 4.4 of the following Chapter 4.

For the discussion in this section, our aim is to reconstruct the unknown function f from its function values $f\big|_X$ by solving the *linear least squares problem*

$$\min_{s \in \mathcal{R}_{\phi,Y}} \sum_{k=1}^{N} |s(x_k) - f(x_k)|^2, \tag{3.63}$$

where for fixed $\phi \in \mathbf{CPD}_d(m)$, the *recovery space* $\mathcal{R}_{\phi,Y} \equiv \mathcal{R}^1_{\phi,Y}$, see (3.56), is given by

$$\mathcal{R}_{\phi,Y} = \left\{ \sum_{j=1}^{n} c_j \phi(\|\cdot - y_j\|) + p \,:\, p \in \mathcal{P}_m^d, \sum_{j=1}^{n} c_j q(y_j) = 0 \text{ for all } q \in \mathcal{P}_m^d \right\}.$$

Note that the form of any $s \in \mathcal{R}_{\phi,Y}$ is similar to that of the interpolant in (3.2). But due to the small size of Y (relative to X), the number of coefficients of s is only $n+q$, which is much smaller than the number of coefficients of any interpolant in (3.2), i.e., $n+q \ll N+q$. Therefore, the above approximation problem (3.63) has a much smaller complexity than the interpolation problem (3.1).

Strictly speaking, the above optimization problem (3.63) is a *linear least squares problem with linear equality constraints.* This is due to the vanishing moment conditions, the linear constraints on $c_1, \ldots, c_n$ in the definition of $\mathcal{R}_{\phi,Y}$. In the following of this section, we show that the problem (3.63) has always a unique solution $s^* \in \mathcal{R}_{\phi,Y}$, provided that Y is $\mathcal{P}_m^d$*-unisolvent*, see the definition in (3.13). This result is stated in Theorem 17 below.

But let us first make some preparations. The unique solution s^* of (3.63) is referred to as the *best approximation* of f w.r.t. the data $f\big|_X$ (in the sense of least squares). Moreover, the expression

$$\min_{s \in \mathcal{R}_{\phi,Y}} \|(s-f)\big|_X\| = \|(s^*-f)\big|_X\| = \left(\sum_{k=1}^{N} |s^*(x_k) - f(x_k)|^2 \right)^{1/2} \tag{3.64}$$

is called the *least squares error.* The solution s^* can be computed by using standard techniques from numerical linear algebra [14, 111].

In order to explain the relevant details on the least squares approximation scheme, we assume, without loss of generality, that the points in X are ordered, such that the points from $Y \subset X$ come first, i.e., $y_k = x_k$, $1 \le k \le n$, and we let $Z = X \setminus Y$.

Moreover, we use the abbreviation

$$A_{\phi,X,Y} = (\phi(\|x_k - y_j\|))_{\substack{1\le k\le N\\ 1\le j\le n}} \in \mathbb{R}^{N\times n}.$$

In particular, $A_{\phi,X,X} \equiv A_{\phi,X}$. Finally, recall the definition of $P_X \in \mathbb{R}^{N\times q}$ in (3.8). The matrices $P_Y \in \mathbb{R}^{n\times q}$ and $P_Z \in \mathbb{R}^{(N-n)\times q}$ are defined accordingly.

Theorem 17. *Let $\phi \in \mathbf{CPD}_d(m)$. Suppose Y is a $\mathcal{P}_m^d$-unisolvent subset of X, $Y \subset X$. Then, the constrained linear least squares problem (3.63) has a unique solution.*

Proof. The linear least squares problem (3.63) can be rewritten as

$$\min_{P_Y^T c=0} \|Bb - f|_X\|, \tag{3.65}$$

where $b = [c^T, d^T]^T \in \mathbb{R}^{n+q}$, $c = (c_1, \ldots, c_n)^T \in \mathbb{R}^n$ and $d = (d_\alpha)_{|\alpha|<m} \in \mathbb{R}^q$, is the coefficient vector of

$$s(x) = \sum_{j=1}^{n} c_j \phi(\|x - y_j\|) + \sum_{|\alpha|<m} d_\alpha x^\alpha \in \mathcal{R}_{\phi,Y}.$$

Moreover, the entries of the matrix

$$B = \left[A_{\phi,X,Y} \middle| P_X\right] \in \mathbb{R}^{N\times(n+q)}$$

are the point evaluations of the basis functions $\phi(\|\cdot - y_j\|)$, $1 \le j \le n$, and x^α, $|\alpha| < m$, at the points in X, respectively. Now note that the constraints $P_Y^T c = 0$ in (3.65) can be expanded as $Cb = 0$, where

$$C = \left[P_Y^T | 0\right] \in \mathbb{R}^{q\times(n+q)}.$$

According to [111, Chapter 20, Theorem 20.9], the problem (3.65) has a solution provided that the matrix C has full rank q. Moreover, under the assumption rank$(C) = q$ a solution of (3.65) is unique, if and only if the augmented matrix

$$D = \begin{bmatrix} C \\ B \end{bmatrix} \in \mathbb{R}^{(N+q)\times(n+q)}$$

has full rank, i.e., rank$(D) = n + q$. But by splitting the matrix D as

$$D = \left[\begin{array}{cc} P_Y^T & 0 \\ A_{\phi,Y} & P_Y \\ \hline A_{\phi,Z,Y} & P_Z \end{array}\right]$$

and by using Theorem 2, we immediately see that rank$(D) = n + q$. Finally, C has full rank due to the injectivity of P_Y which is a direct consequence from our assumption on Y. □

Solving (3.65) by direct elimination requires according to [111] a partitioning of

$$D = \left[\begin{array}{c|c} C_1 & C_2 \\ \hline B_1 & B_2 \end{array}\right] \qquad (3.66)$$

with $C_1 \in \mathbb{R}^{q\times q}, C_2 \in \mathbb{R}^{q\times n}, B_1 \in \mathbb{R}^{N\times q}, B_2 \in \mathbb{R}^{N\times n}$, and

$$b = \begin{bmatrix} b_1 \\ b_2 \end{bmatrix}$$

where $b_1 \in \mathbb{R}^q, b_2 \in \mathbb{R}^n$.

Without loss of generality, we assume that C_1 is nonsingular which is equivalent to requiring that the set $Y_q = \{y_1, \ldots, y_q\} \subset Y$ is $\mathcal{P}_m^d$-unisolvent. Since Y itself is assumed to be $\mathcal{P}_m^d$-unisolvent, it follows that Y must contain a $\mathcal{P}_m^d$-unisolvent subset of size q.

Due to the given constraints $Cb = 0$, we find that $b_1 = -C_1^{-1}C_2b_2$, and therefore by using (3.66) we have

$$\|Bb - f|_X\| = \|\tilde{B}_2b_2 - f|_X\|$$

with $\tilde{B}_2 = B_2 - B_1C_1^{-1}C_2$. Consequently, the computation of the solution of (3.65) can be reduced to solving the unconstrained linear least squares problem

$$\min_{b_2\in\mathbb{R}^n} \|\tilde{B}_2b_2 - f|_X\|. \qquad (3.67)$$

A practical way for computing the solution of (3.67) and thus of (3.65) dates back to Bjørck and Golub [15] (see also [111, Chapter 21]). The starting point in [15] is a decomposition of $C = Q_1^T[\tilde{C}_1, \tilde{C}_2]$, where $Q_1 \in \mathbb{R}^{q\times q}$ is orthogonal and $\tilde{C}_1 \in \mathbb{R}^{q\times q}$ is upper triangular. Using the identity

$$\tilde{B}_2 = B_2 - (B_1\tilde{C}_1^{-1})(Q_1C_2) = B_2 - \tilde{B}_1\tilde{C}_2,$$

the computation of $\tilde{B}_2$ requires solving the triangular system $\tilde{B}_1\tilde{C}_1 = B_1$ for the determination of $\tilde{B}_1 \in \mathbb{R}^{N\times q}$. For solving (3.67) it remains to decompose the matrix $\tilde{B}_2 \in \mathbb{R}^{N\times n}$ by finding an orthogonal matrix $Q_2 \in \mathbb{R}^{N\times N}$ such that

$$Q_2 \cdot [\tilde{B}_2 \,|\, f|_X] = \left[\begin{array}{c|c} \hat{B}_2 & g_1 \\ \hline 0 & g_2 \end{array}\right] \in \mathbb{R}^{N\times(n+1)},$$

where $\hat{B}_2 \in \mathbb{R}^{n\times n}$ is upper triangular, and $g_1 \in \mathbb{R}^n, g_2 \in \mathbb{R}^{N-n}$. The computation of the solution b of (3.65) can then be accomplished by solving the two triangular systems

$$\begin{aligned} \hat{B}_2b_2 &= g_1 \\ \tilde{C}_1b_1 &= -\tilde{C}_2b_2 \end{aligned}$$

one after the other.

We finally remark that the least squares error in (3.64) can obviously be bounded from above by using the available pointwise error estimates of the interpolation scheme, as discussed in Section 3.5. Indeed, this is due to the following simple observation, where $s_{f,Y} \in \mathcal{R}_{\phi,Y}$ denotes the interpolant to f at Y satisfying $s_{f,Y}\big|_Y = f\big|_Y$.

$$\begin{aligned}
\|(s^* - f)\big|_X\|^2 &= \sum_{k=1}^{N} |s^*(x_k) - f(x_k)|^2 \\
&\leq \sum_{k=1}^{N} |s_{f,Y}(x_k) - f(x_k)|^2 \\
&= \sum_{k=n+1}^{N} |s_{f,Y}(x_k) - f(x_k)|^2 \\
&\leq C \cdot |f|_\phi^2 \sum_{z \in Z} F_\phi(h_{\varrho,Y}(z)).
\end{aligned}$$

This immediately leads us to the bound

$$\|(s^* - f)\big|_X\|^2 \leq (N - n) \cdot C \cdot |f|_\phi^2 \cdot F_\phi(h_{\varrho,Y,Z})$$

on the least squares error, where $h_{\varrho,Y,Z} = \max_{z \in Z} h_{\varrho,Y}(z)$.

4 Thinning Algorithms

Thinning algorithms are greedy point removal schemes for scattered data, where the points are recursively removed according to some specific removal criterion. This yields a hierarchy of the input data, which is used for building a multiresolution approximation of a model object, a mathematical function. In general, thinning algorithms are therefore useful tools for model simplification and data reduction.

This chapter concerns the construction and efficient implementation of thinning algorithms for the special case of bivariate scattered data, with emphasis on specific applications. We basically distinguish between two different types of thinning algorithms, *non-adaptive thinning* and *adaptive thinning*. Non-adaptive thinning only depends on the point locations of the input planar point set, whereas in adaptive thinning also function values, sampled at the input points, are used for the point removal.

Adaptive thinning is used in combination with linear splines over Delaunay triangulations, in order to create a multiresolution approximation of the sampled bivariate function. In contrast to adaptive thinning, non-adaptive thinning typically generates a hierarchy of uniformly distributed subsets from the input data. Such data hierarchies, generated by non-adaptive thinning, are used in the next Chapter 5 for the purpose of multilevel interpolation by radial basis functions.

This chapter reviews our previous and current work on thinning algorithms as follows. Section 4.1 first provides a preliminary discussion on thinning algorithms and related concepts, before a generic formulation of thinning is given in Section 4.2. Section 4.3 is then devoted to non-adaptive thinning [74]. This is through Section 4.4 followed by a discussion on *progressive scattered data filtering* [101], which is a combination of non-adaptive thinning and a postprocessing local optimization procedure, termed *exchange*.

Then, in Section 4.5, adaptive thinning algorithms [55] are developed. The utility of adaptive thinning is shown for two different applications, terrain modelling and image compression [40, 42]. Image compression is discussed in detail in Section 4.6, where also the important issue of efficient coding [41] is addressed. The construction in Section 4.6 yields a complete image compression scheme, which is shown to be competitive to the well-established wavelet-based compression scheme SPIHT.

4.1 Preliminary Remarks

Conceptually, thinning is a simple approach to generating a multiresolution representation of a triangulated set of points in the plane. The idea of thinning is to recursively remove points, one by one, retriangulating at each step the reduced point set. In [74], where the term *thinning* is introduced, several different strategies, *thinning algorithms*, for point removal are suggested. These thinning algorithms tend to favour well-distributed planar point sets.

Provided one always uses Delaunay triangulations, the only change in the topology at each point removal is *local*. Indeed, at any step one merely needs to fill the whole left by the point removal, i.e., by the removal of its adjacent triangles. This is done by the *retriangulation* of the *cell* $C(y)$ of the removed (vertex) point y, see Figure 4.1. Recall that the cell $C(y)$ of a vertex y in a triangulation $\mathcal{T}_Y$ is the union of all triangles in $\mathcal{T}_Y$ which contain y as a vertex.

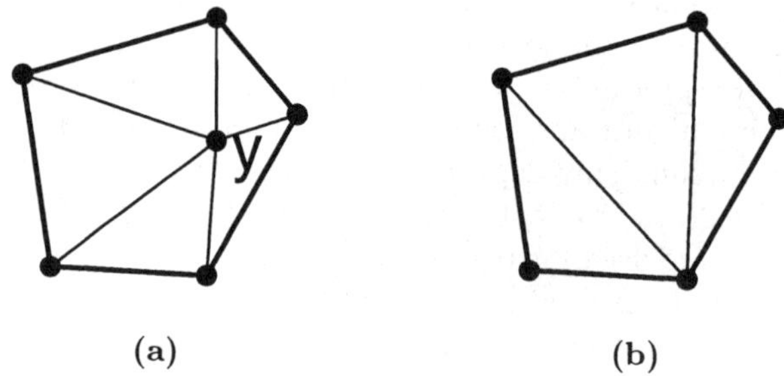

Fig. 4.1. Removal of the node y, and retriangulation of its cell $C(y)$. The five triangles of the cell $C(y)$ in **(a)** are replaced by the three triangles in **(b)**.

Yet it seems that for reducing the approximation error, especially when working with piecewise linear interpolation over triangulated subsets, a better strategy, instead of using Delaunay triangulations, would be that of data-dependent triangulations, as discussed in Section 2.4. But one should note that, when working with data-dependent triangulations, the topological changes required for decremental triangulation are not guaranteed to be local. This may lead to an increased computational overhead during the thinning.

In this chapter, we are concerned with both *non-adaptive thinning* [74], and *adaptive thinning* [55]. In contrast to non-adaptive thinning, adaptive thinning does also take the function values at the 2D points into account. For an up-to-date survey on adaptive thinning, we refer to [40].

The idea of thinning triangulated scattered data is not new. Thinning is only one of several mesh simplification methods, which are more commonly referred to as *mesh decimation* or *mesh simplification* in the literature.

Heckbert and Garland [86] give an extensive survey of simplification methods both for terrain models (triangulated scattered data in the plane) and free form models (manifold surfaces represented by 3D triangle meshes). Specific *mesh decimation* algorithms, other than thinning, include techniques like edge-collapse, half-edge collapse, and vertex collapse. For a recent tutorial on these methods, see the paper [80] by Gotsman, Gumhold, and Kobbelt.

We remark that there exist further alternative methods for generating hierarchical triangulations. One common concept is given by *insertion algorithms*, where points are successively added to a coarse triangulation. Therefore, insertion can be viewed as the inverse of thinning. Both techniques, thinning and insertion, essentially require careful management of the data structure needed to represent a hierarchical sequence of triangulations, termed *multi-triangulation*. For a recent account on the state-of-the-art in this field, we recommend the tutorial paper [38] by De Floriani and Magillo.

4.2 Generic Formulation

In the following discussion of this chapter, $X \subset \mathbb{R}^2$ denotes a finite scattered point set of size $|X| = N < \infty$. A generic formulation of thinning on X is given by the following algorithm, where n is the number of removal steps.

Algorithm 16 (Thinning).

INPUT: *X with $|X| = N$, and $n \in \{1, \ldots, N-1\}$;*

(1) *Let $X_N = X$;*
(2) FOR $k = 1, \ldots, n$
  (2a) *Locate a* `removable` *point $x \in X_{N-k+1}$;*
  (2b) *Let $X_{N-k} = X_{N-k+1} \setminus x$;*

OUTPUT: *$X_{N-n} \subset X$, of size $|X_{N-n}| = N - n$.*

In order to select a specific thinning strategy, it remains to give a definition for a removable point in step **(2a)** of the Algorithm 16. But this should depend on the specific requirements of the underlying application.

On this occasion, we remark that we consider applying *adaptive* thinning, in combination with linear splines over Delaunay triangulations, to two different applications, *terrain modelling* (Subsection 4.5.5) and *image compression* (Section 4.6). By the different requirements of these two applications, *customized* removal criteria, for step **(2a)** of Algorithm 16, need to be developed. The construction of suitable adaptive *anticipated error measures*, on which the different removal criteria rely, is done in Subsection 4.5.4, for terrain modelling, and in Subsection 4.6.2, for image compression.

Note that during the performance of the above thinning algorithm a *nested* sequence

$$X_{N-n} \subset X_{N-n+1} \subset \cdots \subset X_{N-1} \subset X_N = X \tag{4.1}$$

of subsets of X is computed. This yields also an ordering of the points in X. Indeed, note that two consecutive sets $X_{N-k} \subset X_{N-k+1}$, $1 \leq k \leq n$, in (4.1) differ about only one point. This difference is the point $x_k = X_{N-k+1} \setminus X_{N-k}$ which is removed in the k-th step in **(2b)** of the thinning algorithm.

For the subsequent discussion, it is convenient to associate with any thinning algorithm a *thinning operator* T. The operation of T on any non-empty subset $Y \subset X$, is defined by $T(Y) = Y \setminus y$ for one unique $y \in Y$, so by the action of T on Y, the point y is removed from Y. Therefore, any subset X_{N-n} output by Algorithm 16 can be written as $X_{N-n} = T^n(X)$, where $T^n = T \circ \cdots \circ T$ denotes the n-fold composition of T. Hence, the data hierarchy (4.1) can be expressed as

$$T^n(X) \subset T^{n-1}(X) \subset \cdots \subset T(X) \subset T^0(X) = X, \tag{4.2}$$

where T^0 denotes the identity.

In our applications, we use nested sequences of the above form (4.2) in order to construct a *coarser* data hierarchy

$$X_{N-n_{L-1}} \subset X_{N-n_{L-2}} \subset \cdots \subset X_{N-n_2} \subset X_{N-n_1} \subset X \tag{4.3}$$

comprising L *data levels*, where $n_1, \ldots, n_{L-1}$ is a strictly increasing sequence of *breakpoints*. Note that the data hierarchy in (4.3) already yields a multiresolution representation for the given set X. But the selection of the breakpoints therein requires care. Details on this will be deferred to later in this chapter.

4.3 Non-Adaptive Thinning

Non-adaptive thinning concerns the construction of an entire sequence

$$X_{N-n} \subset X_{N-n+1} \subset \cdots \subset X_{N-1} \subset X_N = X \tag{4.4}$$

of nested subsets, with increasing sizes $|X_{N-n}| = N - n$, such that for each $Y \equiv X_{N-n} \subset X$ its *covering radius*

$$r_{Y,X} = \max_{x \in X} d_Y(x) \tag{4.5}$$

on X is small, where

$$d_Y(x) = \min_{y \in Y} \|y - x\|$$

denotes the Euclidean distance between the point $x \in X$ and the subset $Y \subset X$, see (2.4). The progressive construction of such a sequence is accomplished by using one *greedy* thinning operator T_*, so that each of the subsets X_{N-n} in (4.4) can be written as $X_{N-n} = T_*(X_{N-n+1})$, i.e., $X_{N-n} = T_*^n(X)$, for $1 \leq n \leq N-1$, and thus satisfies $|X_{N-n}| = N - n$.

4.3.1 The k-center Problem

Before we proceed with explaining details on non-adaptive thinning, which is the subject of most of this section, let us first make a few preliminary remarks. Observe that for the above purpose one ideally wants to pick, for any n, one subset $Y \equiv X^*_{N-n} \subset X$, with (small) size $|X^*_{N-n}| = N - n$, which is *optimal* by minimizing the covering radius $r_{Y,X}$ among all subsets $Y \subset X$ of equal size, so that

$$r^*_n = r_{X^*_{N-n},X} = \min_{\substack{Y \subset X \\ |Y|=N-n}} r_{Y,X}. \tag{4.6}$$

The problem of finding an algorithm which outputs for any possible input pair (X, n), $1 \leq n < |X|$, such an optimal subset X^*_{N-n} satisfying (4.6) is one particular instance of the *k-center problem*. In the more general d-dimensional setting, the norm $\|\cdot\|$ may be replaced by any arbitrary metric on $\mathbb{R}^d$.

But the k-center problem is, due to Kariv and Hakimi [107], NP-hard. Moreover, the problem of finding an *α-approximation algorithm*, $\alpha \geq 1$, for the k-center problem which outputs for any input pair (X, n), $1 \leq n < |X|$, a subset $X_{N-n} \subset X$ of size $|X_{N-n}| = N - n$ satisfying

$$r_{X_{N-n},X} \leq \alpha \cdot r^*_n \tag{4.7}$$

is for any $\alpha < 2$ NP-complete. Hochbaum and Shmoys [89] were the first to provide a 2-approximation algorithm (i.e., $\alpha = 2$) for the k-center problem, which is best possible unless P=NP. For a comprehensive discussion on the k-center problem we refer to the textbook [88, Section 9.4.1] and the survey paper [164], where the Hochbaum-Shmoys algorithm is explained.

In contrast to the situation in the k-center problem, non-adaptive thinning does not work with a beforehand selection for $N - n$, the size of the output $X_{N-n} \subset X$. Instead of this, the algorithms picks one *good* subset X_{N-n} at run time, by selecting one suitable breakpoint $n \equiv n_j$ in (4.3). For the purpose of controlling the covering radius $r_{X_{N-n},X}$, this selection relies on *adaptive* bounds of the form

$$r_{X_{N-n},X} \leq \alpha_{X_{N-n},X} \cdot r^*_n, \tag{4.8}$$

where $\alpha_{X_{N-n},X} = r_{X_{N-n},X}/\sigma_n$ denotes the *quality index* of X_{N-n}, and the numbers σ_n solely depend on the distribution of the points in X. The adaptive bounds in (4.8) are proven in the following Subsection 4.3.2.

Note that the upper bound on $r_{X_{N-n},X}$ in (4.8) looks similar to the one in (4.7). However, while $\alpha_{X_{N-n},X}$ in (4.8) depends on both X and $X_{N-n} \subset X$, the universal constant α in (4.7) does not even depend on X. In fact, the sequence of numbers $\alpha_{X_{N-n},X}$, recorded at run time, helps us to control the *relative* deviation

$$\left| \frac{r_{X_{N-n},X} - r^*_n}{r^*_n} \right| \leq \alpha_{X_{N-n},X} - 1, \tag{4.9}$$

between any current covering radius $r_{X_{N-n},X}$ and the optimal value r_n^*. Note that the bound (4.9) follows directly from (4.8).

Whenever $\alpha_{X_{N-n},X}$ is close to one, this then indicates that the set X_{N-n} is close to one optimal set of equal size $|X_{N-n}| = N - n$. In our applications in [94, 96], this turns out to be a useful criterion for the selection of *good* subsets from X.

The numerical results in Subsection 4.4.4 show how greedy (non-adaptive) thinning, to be explained in Subsection 4.3.3, and the related *scattered data filtering*, being subject of Section 4.4, performs in comparison with possible α-approximation algorithms for the k-center problem.

4.3.2 Adaptive Bounds on the Covering Radii

In this section, adaptive bounds on the covering radius $r_{Y,X}$, for any $Y \subset X$, are proven. To this end, assume without loss of generality that the points in $X = \{x_1, \ldots, x_N\}$ are ordered such that their *significances*

$$\sigma(x) = d_{X \setminus x}(x), \qquad \text{for } x \in X, \tag{4.10}$$

are increasing, i.e.,

$$\sigma(x_1) \leq \sigma(x_2) \leq \cdots \leq \sigma(x_N). \tag{4.11}$$

Note that for any $x \in X$ its significance $\sigma(x)$ in (4.10) is the Euclidean distance to its nearest neighbour in X. Hence, according to the above assumption (4.11) on the ordering of the points in X, the value $\sigma(x_1)$ yields the minimal distance between two points in X. In fact, since this minimum is attained by at least two points in X, we have $\sigma(x_1) = \sigma(x_2)$.

For notational simplicity, we let $\sigma_n = \sigma(x_n)$ for $1 \leq n \leq N$. Moreover, for any $Y \subset X$ of size $|Y| = N - n$, we let $Z = X \setminus Y$ denote the difference set, whose size is then $|Z| = n$. Starting point of the subsequent discussion is the following lower bound on the covering radius $r_{Y,X}$ for $Y \subset X$.

Theorem 18. *For any $Y \subset X$ of size $|Y| = N - n$ the inequality*

$$\sigma_n \leq r_{Y,X} \tag{4.12}$$

holds.

Proof. Since for any $z \in Z = X \setminus Y$ the inequality

$$d_{X \setminus Z}(z) \geq d_{X \setminus z}(z) = \sigma(z)$$

holds, we conclude

$$r_{Y,X} = r_{X \setminus Z,X} = \max_{x \in X} d_{X \setminus Z}(x) = \max_{z \in Z} d_{X \setminus Z}(z) \geq \max_{z \in Z} \sigma(z). \tag{4.13}$$

By our assumption (4.11) on the ordering of the points in X and by $|Z| = n$, it follows that

$$\max_{z \in Z} \sigma(z) \geq \sigma(x_n) = \sigma_n$$

which completes, by using (4.13), our proof. □

Note that the above inequality (4.12) holds in particular for any optimal set $X^*_{N-n} \subset X$ of size $|X^*_{N-n}| = N-n$ satisfying $r_{X^*_{N-n},X} = r^*_n$, which yields $\sigma_n \leq r^*_n$ for $n = 1, \ldots, N-1$. This immediately implies

$$r_{Y,X} = \alpha_{Y,X} \cdot \sigma_n \leq \alpha_{Y,X} \cdot r^*_n,$$

where we let $\alpha_{Y,X} = r_{Y,X}/\sigma_n$. This is the adaptive upper bound (4.8) stated in the previous Subsection 4.3.1. In summary, we draw the following conclusion from Theorem 18.

Corollary 3. *For any $Y \subset X$ of size $|Y| = N - n$ the inequalities*

$$\sigma_n \leq r^*_n \leq r_{Y,X} \leq \alpha_{Y,X} \cdot r^*_n \tag{4.14}$$

hold, where $\alpha_{Y,X} = r_{Y,X}/\sigma_n \geq 1$. □

In situations where the optimal value r^*_n is known, the following observation may help to construct an optimal subset by using the initial significances of the points in X.

Theorem 19. *Suppose $Y \subset X$ is an optimal subset of size $|Y| = N - n$. Then, this implies $\sigma(z) \leq r^*_n$ for all $z \in Z = X \setminus Y$.*

Proof. Note that every point $z \in Z$ satisfies

$$\sigma(z) = d_{X \setminus z}(z) \leq d_{X \setminus Z}(z) = d_Y(z) \leq r_{Y,X}. \tag{4.15}$$

Moreover, since Y is optimal, we have $r_{Y,X} = r^*_n$. This in combination with (4.15) implies $\sigma(z) \leq r^*_n$ for every $z \in Z$, as stated. □

Note that the above characterization implies that any optimal $Y^* \subset X$ of size $N - n$ is necessarily a superset of $X \setminus \{x \in X : \sigma(x) \leq r^*_n\}$. We come back to this point in Subsection 4.3.4.

4.3.3 Greedy Thinning

Greedy algorithms are known as efficient and effective methods of dynamic programming for solving optimization problems. Greedy algorithms typically go through a sequence of steps, where for each step a choice is made that looks best at the moment. For a general introduction to greedy algorithms we recommend the textbook [33, Chapter 16].

In our particular situation, a greedy thinning algorithm is one where at each step one point is removed, such that the resulting covering radius is minimal among all other possible point removals. This leads us to the following definition for a removable point in step **(2a)** of Algorithm 16.

Definition 13. *For any $Y \subset X$ with $|Y| \geq 2$, a point $y^* \in Y$ is said to be* `removable` *from Y, iff y^* minimizes the covering radius $r_{Y \setminus y, X}$ among all points in Y, i.e.,*

$$r_{Y \setminus y^*, X} = \min_{y \in Y} r_{Y \setminus y, X}.$$

We remark that this definition for a removable point is different from those used in [72, 74, 96], where a removable point is one which minimizes the distance to its nearest neighbour in the *current* subset Y. In contrast to this, the removal criterion of Definition 13 depends also on the points in $Z = X \setminus Y$ which have already been removed in previous steps. This idea is also favourably used in the recent paper [55].

At first sight, the task of locating a removable point may look costly. The computation can, however, be facilitated by using the following characterization for removable points, which works with Voronoi diagrams. To this end, recall from the discussion in Section 2.3 that for any finite point set Y and $y \in Y$ the convex polyhedron

$$V_Y(y) = \left\{x \in \mathbb{R}^2 \; : \; d_Y(x) = \|x - y\|\right\}$$

denotes the *Voronoi tile* of y w.r.t. Y, comprising all points in the plane whose nearest neighbour in Y is y.

Theorem 20. *Let $Y \subset X$ with $|Y| \geq 2$. Every point $y \in Y$ which minimizes the* `local covering radius`

$$r(y) = r_{Y \setminus y, X \cap V_Y(y)} \tag{4.16}$$

among all points in Y is removable from Y.

Proof. Let $Z = X \setminus Y$. Note that

$$\begin{aligned} r_{Y \setminus y, X} &= \max_{z \in Z \cup y} d_{Y \setminus y}(z) \\ &= \max\left(\max_{z \in Z \setminus V_Y(y)} d_{Y \setminus y}(z), \max_{z \in Z \cap V_Y(y)} d_{Y \setminus y}(z), d_{Y \setminus y}(y)\right) \\ &= \max\left(\max_{z \in Z \setminus V_Y(y)} d_Y(z), \max_{z \in Z \cap V_Y(y)} d_{Y \setminus y}(z), d_{Y \setminus y}(y)\right). \end{aligned}$$

Since $d_{Y \setminus y}(z) \geq d_Y(z)$ for all $z \in Z \cap V_Y(y)$, this implies

$$r_{Y \setminus y, X} = \max\left(\max_{z \in Z} d_Y(z), \max_{z \in Z \cap V_Y(y)} d_{Y \setminus y}(z), d_{Y \setminus y}(y)\right).$$

Moreover, since

$$\max\left(\max_{z \in Z \cap V_Y(y)} d_{Y \setminus y}(z), d_{Y \setminus y}(y)\right) = \max_{z \in (Z \cap V_Y(y)) \cup y} d_{Y \setminus y}(z) = r_{Y \setminus y, X \cap V_Y(y)}$$

and $r_{Y,X} = \max_{z \in Z} d_Y(z)$, we obtain

$$r_{Y \setminus y, X} = \max\left(r_{Y,X}, r(y)\right). \tag{4.17}$$

Therefore, $r_{Y \setminus y, X} \leq r_{Y \setminus \tilde{y}, X}$, whenever $r(y) \leq r(\tilde{y})$ for any $y, \tilde{y} \in Y$, which completes our proof. □

4.3.4 Localization of Optimal Subsets

In this subsection, one useful (theoretical) property of greedy thinning is discussed. This is concerning the localization of optimal subsets during the thinning process. To this end, we use the notation $T_*^n(X) \subset X$ for a subset output by greedy thinning after n point removals. In particular, we have $|T_*^n(X)| = N - n$ for the size of $T_*^n(X)$, and so by Corollary 3 in Subsection 4.3.2 we obtain for any $n \in \{1, \ldots, N-1\}$ the adaptive bounds

$$\sigma_n \leq r_n^* \leq r_{T_*^n(X),X} \leq \alpha_{T_*^n(X),X} \cdot r_n^*$$

for the covering radius of $T_*^n(X)$ on X, where $\alpha_{T_*^n(X),X} = r_{T_*^n(X),X}/\sigma_n \geq 1$.

Now, if $\alpha_{T_*^n(X),X} = 1$, this then would directly imply that the subset $T_*^n(X)$ is optimal with satisfying $r_{T_*^n(X),X} = \sigma_n = r_n^*$. For instance, at the first point removal (i.e., when $n = 1$) greedy thinning returns the optimal subset $T_*(X) = X \setminus x^*$, with x^* being some removable point from X, satisfying

$$r_{T_*(X),X} = r_{X \setminus x^*, X} = d_{X \setminus x^*}(x^*) = \sigma(x^*) = \sigma_1.$$

But for general n (and general X), it is not true that σ_n coincides with $r_{T_*^n(X),X}$. This leads us to the following definition.

Definition 14. *For given X of size $|X| = N$, an index n, $1 \leq n < N$, is said to be an* `optimal breakpoint` *for X, iff there is one $Y \subset X$ of size $|Y| = N - n$ satisfying $r_{Y,X} = \sigma_n$.*

Hence, for any input X, $n = 1$ is always an optimal breakpoint. Indeed, in this case $Y = T_*(X)$ satisfies $r_{Y,X} = \sigma_1$. But in general, i.e., for $n > 1$, it is not necessarily true that n is an optimal breakpoint for X. Nevertheless, whenever any n is an optimal breakpoint for X, we can show that the subset $T_*^n(X) \subset X$ generated by greedy thinning is the *unique* optimum satisfying $T_*^n(X) = \sigma_n$, provided that $\sigma_n < \sigma_{n+1}$.

Theorem 21. *Suppose n is an optimal breakpoint for X. If $\sigma_n < \sigma_{n+1}$, then the subset*

$$X_{N-n}^* = X \setminus \{x_1, \ldots, x_n\} \subset X$$

is optimal by satisfying $r_{X_{N-n}^*,X} = \sigma_n$. Moreover, X_{N-n}^* is the unique minimizer of the covering radius $r_{Y,X}$ among all subsets $Y \subset X$ of equal size $|Y| = N - n$.

Proof. Since n is an optimal breakpoint in X, there is at least one optimal subset $Y \subset X$ of size $|Y| = N - n$ satisfying $r_{Y,X} = \sigma_n$. Let $Z = X \setminus Y$. Then, due to Theorem 19, this implies

$$\sigma(z) \leq \sigma_n \tag{4.18}$$

for every $z \in Z$. But since $\sigma_n < \sigma_{n+1}$ and by (4.11), the condition (4.18) is only satisfied by the points in the set $X_n = \{x_1, \ldots, x_n\}$, and so $Z \subset X_n$. But $|Z| = |X_n| = n$, and therefore $Z = X_n$, which implies $Y = X \setminus X_n$. □

Corollary 4. *Suppose n is an optimal breakpoint for X, and $\sigma_n < \sigma_{n+1}$. Then, the subset $T_*^n(X) \subset X$ output by greedy thinning is optimal by satisfying $r_{T_*^n(X)} = \sigma_n$. Moreover, $T_*^n(X)$ is the unique minimizer of the covering radius $r_{Y,X}$ among all subsets $Y \subset X$ of equal size $|Y| = N - n$.*

Proof. Due to the above Theorem 21, we need to show that $T_*^n(X) = X \setminus X_n$, where $X_n = \{x_1, \ldots, x_n\}$. We prove this by induction. Since $r_{X \setminus x, X} = \sigma(x)$ and due to the assumption $\sigma_n < \sigma_{n+1}$, greedy thinning removes one point from X_n in its first step, i.e., $T_*(X) = X \setminus x^*$ for some $x^* \in X_n$.

Now suppose, for any $1 \leq k < n$, that $T_*^k(X) = X \setminus Z$ holds with $Z \subset X_n$. Then, on the one hand, for every $z \in X_n \setminus Z \subset X_n$ we have

$$r_{T_*^k(X) \setminus z, X} = r_{X \setminus (Z \cup z), X} \leq r_{X \setminus X_n, Y} = \sigma_n. \tag{4.19}$$

Indeed, this is due to the monotonicity of the covering radius, i.e.,

$$r_{Y,X} \leq r_{\tilde{Y},X}, \qquad \text{for all } Y, \tilde{Y} \text{ with } \tilde{Y} \subset Y \subset X. \tag{4.20}$$

On the other hand, for every $x \in X \setminus X_n = \{x_{n+1}, \ldots, x_N\}$ we have

$$d_{T_*^k(X) \setminus x}(x) = d_{(X \setminus Z) \setminus x}(x) \geq d_{X \setminus x}(x) = \sigma(x) > \sigma_n,$$

and so $r_{T_*^k(X) \setminus x, X} > \sigma_n$. This in combination with (4.19) shows that

$$r_{T_*^k(X) \setminus z, X} < r_{T_*^k(X) \setminus x, X}, \qquad \text{for all } z \in X_n \setminus Z, x \in X \setminus X_n.$$

Therefore, greedy thinning removes one point from $X_n \setminus Z \subset X_n$ in its next step. After n removals, we have $T_*^n(X) = X \setminus X_n$ as desired. □

4.3.5 Implementation of Greedy Thinning

Now let us turn to the important point of efficiently implementing greedy thinning algorithms. To this end, recall the Definition 13 for a removable point, and the characterization in Theorem 20. The latter says that a point $y^* \in Y$, which minimizes the local covering radius $r(y)$ among all points in $Y \subset X$, is removable from Y.

Following along the lines of the discussion in the previous papers [74, 55], during the performance of the greedy thinning algorithm, the points of the current set Y are stored in a heap. Recall that a heap is a binary tree which can be used for the maintenance of a priority queue (see Section 2.5 for details). Each node $y \in Y$ in the heap bears its local covering radius $r(y)$ in (4.16) as its *significance value.*

The nodes in the heap are sorted, such that the significance of a node is *smaller* than the significances of its two children. Hence, due to the heap condition, the root of the heap contains a least significant node, and thus a removable point. Therefore, the removal of a removable point in steps **(2a)** and **(2b)** of the thinning Algorithm 16 can be performed by *popping* the root of the heap by using the function `remove`, Algorithm 10 in Subsection 2.5.4. Recall that building the initial heap costs $\mathcal{O}(N \log N)$ operations, where $N = |X|$ is the size of the input point set X. Likewise, building the initial Voronoi diagram of X costs $\mathcal{O}(N \log N)$ operations. But each point removal requires also a subsequent update of both the employed data structures, the heap and the Voronoi diagram, and the significance values of the nodes in the heap.

More precise, the following steps are performed when removing one removable point by greedy thinning.

(T1) Pop the root y^* from the heap and update the heap.
(T2) Remove y^* from the Voronoi diagram. Update the Voronoi diagram in order to obtain the Voronoi diagram of the point set $Y \setminus y^*$.
(T3) Let $Y = Y \setminus y^*$ and so $Z = Z \cup y^*$.
(T4) Update the local covering radii of the Voronoi neighbours of y^* in Y, whose Voronoi tiles were changed by the update in step **(T2)**. Update the positions of these points in the heap.

In addition, during the performance of greedy thinning, each $z \in Z$ is attached to a Voronoi tile containing z. Thus in step **(T2)**, by the removal of y^*, the points in $Z \cap V_Y(y^*)$ and y^* itself need to be *reattached* to new Voronoi tiles $V_{Y \setminus y^*}(\cdot)$ of Voronoi neighbours of y^*. These (re)attachments facilitate the required updates of the local covering radii in step **(T4)**.

We remark that the updates in the above steps **(T2)** and **(T4)** require merely local operations on the Voronoi diagram. In fact, each of these two steps cost $\mathcal{O}(1)$ operations, provided that we have $|Z \cap V_Y(y^*)| = \mathcal{O}(1)$ for the number of points in Z that need to be reattached to new Voronoi tiles. Moreover, each update on the heap (in steps **(T1)** and **(T4)**) costs at most $\mathcal{O}(\log N)$ operations, where $|Y| = N - n$ is the number of nodes in the heap (see Subsection 2.5.4). Altogether, this shows that each removal step of greedy thinning costs at most $\mathcal{O}(\log N)$ operations, provided that $|Z \cap V_Y(y^*)| = \mathcal{O}(1)$.

But in fact, under the assumption that the $|Z| = n$ points in Z are uniformly distributed over the $N - n$ Voronoi tiles, we have $|Z \cap V_Y(y^*)| \approx n/(N - n)$ for the number of points in $Z \cap V_Y(y^*)$. Therefore, the number of operations in each of the steps **(T2)** and **(T4)** are of order $n/(N - n)$. By summing up the computational costs for **(T1)**-**(T4)** over $N - 1$ point removals, this shows that the total costs for generating the entire data hierarchy in (4.4) by using greedy thinning are at most $\mathcal{O}(N \log(N))$. In summary, we obtain the following result concerning the computational costs of greedy thinning.

Theorem 22. *The performance of the thinning Algorithm 16, by using the removal criterion of Definition 13, and according to the steps* **(T1)-(T4)** *requires at most* $\mathcal{O}(N \log N)$ *operations.* □

4.4 Scattered Data Filtering

Scattered data filtering is a progressive data reduction scheme for scattered data. The scheme is a combination of greedy thinning and a postprocessing local optimization procedure, termed *exchange*, to be explained in the following Subsection 4.4.1. The overall aim of applying (progressive) scattered data filtering on X is to generate a sequence $\{X_{N-n}\}_n$ of subsets of X such that the covering radii $r_{X_{N-n},X}$ are small. This is achieved by requiring that every subset $X_{N-n} \subset X$ is locally optimal in X.

Definition 15. *Let* $Y \subset X$ *and* $Z = X \setminus Y \subset X$. *The set* Y *is said to be* `locally optimal` *in* X, *iff there is no pair* $(y, z) \in Y \times Z$ *of points satisfying*

$$r_{Y,X} > r_{(Y \setminus y) \cup z, X}. \tag{4.21}$$

A point pair $(y, z) \in Y \times Z$ *satisfying (4.21) is said to be* `exchangeable` .

Hence, if $Y \subset X$ is locally optimal in X, then the covering radius $r_{Y,X}$ of Y on X cannot be reduced by any single exchange between a point $y \in Y$ and a point z in the difference set $Z = X \setminus Y$. Note that every *globally optimal* subset X^*_{N-n} satisfying $r_{X^*_{N-n},X} = r^*_n$ is also locally optimal.

Now the idea of progressive scattered data filtering is to reduce, for any subset $Y \equiv T^n_*(X)$ output by greedy thinning, its covering radius $r_{Y,X}$ on X by iteratively swapping point pairs between Y and $Z = X \setminus Y$. The latter is the exchange algorithm, Algorithm 17 in Subsection 4.4.1.

4.4.1 Exchange

Exchange outputs, on any given subset $Y \subset X$ and superset X, a locally optimal subset $Y^* \subset X$ of equal size, i.e., $|Y^*| = |Y|$. This is accomplished, according to the following exchange algorithm, by iteratively swapping exchangeable point pairs between Y and the difference set $Z = X \setminus Y$.

Algorithm 17 (Exchange).

INPUT: *$Y \subset X$;*

(1) *Let $Z = X \setminus Y$;*
(2) WHILE *(Y not locally optimal in X)*
  (2a) *Locate an exchangeable pair $(y, z) \in Y \times Z$;*
  (2b) *Let $Y = (Y \setminus y) \cup z$ and $Z = (Z \setminus z) \cup y$;*

OUTPUT: *$Y \subset X$, locally optimal in X.*

Note that the exchange algorithm terminates after finitely many steps. Indeed, this is because the set X is assumed to be finite, and each exchange in step **(2b)** strictly reduces the current (non-negative) covering radius $r_{Y,X}$. By construction, the output set $Y \subset X$ is then locally optimal. A characterization of exchangeable point pairs is provided in Subsection 4.4.2. This yields useful criteria for the efficient localization of such point pairs.

We associate the exchange Algorithm 17 with an exchange operator E, which returns on any given argument $Y \subset X$ a locally optimal subset $E(Y) \subset X$ of equal size $|E(Y)| = |Y|$. Hence, E is a projector onto the set of locally optimal subsets in X.

Moreover, by the composition of a greedy thinning operator T_* and the exchange operator E, we obtain a sequence $\mathcal{F} = \{F_n\}_n$ of *filter operators*, where $F_n = E \circ T_*^n$. This already yields, as desired, by $X_{N-n} = F_n(Z)$ a sequence $\{X_{N-n}\}_n$ of locally optimal subsets with decreasing size $|X_{N-n}| = N - n$ (at increasing n). But in contrast to the data hierarchy in (4.4), the sets X_{N-n} do not necessarily form a *nested* sequence of subsets.

Nevertheless, note that this progressive scattered data filtering scheme can be applied recursively in order to generate a data hierarchy of the form (4.3), and such that the covering radii $r_{X_{N-n_j}, X_{N-n_{j-1}}}$ of two consecutive subsets $X_{N-n_j} \subset X_{N-n_{j-1}}$, $1 < j < L$, are small. This is done by using the recursion

$$X_{N-n_j} = F_{\Delta n_j}(X_{N-n_{j-1}}),$$

for $j = 1, 2, \ldots, L-1$, where we let $n_0 = 0$ and $\Delta n_j = n_j - n_{j-1}$. In fact, this construction is useful for various multilevel approximation schemes, which are discussed in the following Chapter 5.

4.4.2 Characterization of Exchangeable Point Pairs

This subsection is devoted to the characterization of exchangeable point pairs, which includes important computational aspects concerning the efficient implementation of the exchange Algorithm 17. To this end, useful criteria for an *efficient* localization of exchangeable point pairs are proven. Theorem 23 gives a necessary and sufficient characterization for exchangeable point pairs, whereas the Corollaries 5 and 6 provide sufficient criteria, which are useful for the efficient implementation of step (2a) of the exchange Algorithm 17.

For the moment of the discussion in this subsection, $Y \subset X$ denotes a fixed subset of X and we let $Z = X \setminus Y$. Moreover,

$$Z^* = \{z \in Z : d_Y(z) = r_{Y,X}\} \tag{4.22}$$

stands for the set of all $z \in Z$ where the maximum $r_{Y,X}$ is attained. The following theorem yields a necessary and sufficient condition for exchangeable point pairs. The subsequent two corollaries provide sufficient conditions, which are useful for the purpose of quickly locating exchangeable points.

Theorem 23. *A point pair $(\hat{y}, \hat{z}) \in Y \times Z$ is exchangeable, if and only if all of the following three statements are true.*

(a) $r_{Y,X} > d_{Y \cup \hat{z}}(z)$ *for all* $z \in Z^*$*;*
(b) $r_{Y,X} > d_{(Y \setminus \hat{y}) \cup \hat{z}}(\hat{y})$*;*
(c) $r_{Y,X} > d_{(Y \setminus \hat{y}) \cup \hat{z}}(z)$ *for all* $z \in Z \cap V_Y(\hat{y})$*.*

Before we give a rigorous proof of this theorem below, let us pause to make some remarks concerning the plausibility of the three conditions (a),(b), and (c). To this end, recall that the aim of the exchange between $\hat{y}$ and $\hat{z}$ is to (strictly) reduce the current covering radius $r_{Y,X}$. The three requirements (a),(b), and (c) in Theorem 23 are interpreted individually as follows.

(a) By the insertion of $\hat{z}$ into Y, the maximum $r_{Y,X}$ must be (strictly) reduced. This requires that $\hat{z}$ is for any $z \in Z^*$ the *nearest* point in the set $Y \cup \hat{z}$, with satisfying $d_{Y \cup \hat{z}}(z) = \|\hat{z} - z\| < r_{Y,X}$ for all $z \in Z^*$.
(b) By the removal of $\hat{y}$ from Y, the distance between $\hat{y}$ and its nearest point in the modified set $(Y \setminus \hat{y}) \cup \hat{z}$ must not be greater or equal than the current covering radius $r_{Y,X}$. Otherwise, $r_{Y,X}$ won't be reduced, it may even be increased.
(c) Note that the distance between any point $z \in Z$, lying in the Voronoi tile $V_Y(\hat{y})$ of $\hat{y}$, and the set Y is $\|z - \hat{y}\|$. Therefore, for every point $z \in Z \cap V_Y(\hat{y})$, the distance to its nearest point in the modified set $(Y \setminus \hat{y}) \cup \hat{z}$ is to be updated after the removal of $\hat{y}$ from Y. Hence, the distance between every such point $z \in Z \cap V_Y(\hat{y})$ and the modified set $(Y \setminus \hat{y}) \cup \hat{z}$ must not be greater or equal than the current covering radius $r_{Y,X}$. Otherwise, $r_{Y,X}$ won't be reduced, it may even be increased.

Proof of Theorem 23. Suppose all of the three statements (a),(b), and (c) are true. Note that condition (a), together with the definition for Z^*, implies

$$r_{Y,X} > d_{Y \cup \hat{z}}(z) \quad \text{for all } z \in Z. \tag{4.23}$$

Moreover, for any $z \in Z \setminus V_Y(\hat{y})$ we have $d_Y(z) = d_{Y \setminus \hat{y}}(z)$, and therefore

$$d_{Y \cup \hat{z}}(z) = d_{(Y \setminus \hat{y}) \cup \hat{z}}(z) \quad \text{for all } z \in Z \setminus V_Y(\hat{y}).$$

This, in combination with statement (c) and (4.23), implies

$$r_{Y,X} > d_{(Y\setminus\hat{y})\cup\hat{z}}(z) \quad \text{for all } z \in Z. \tag{4.24}$$

By combining (4.24) with condition (b), we find

$$\begin{aligned} r_{Y,X} &> \max\left(\max_{z\in Z\setminus\hat{z}} d_{(Y\setminus\hat{y})\cup\hat{z}}(z), d_{(Y\setminus\hat{y})\cup\hat{z}}(\hat{y})\right) \\ &= \max_{z\in(Z\setminus\hat{z})\cup\hat{y}} d_{(Y\setminus\hat{y})\cup\hat{z}}(z) \\ &= \max_{x\in X} d_{(Y\setminus\hat{y})\cup\hat{z}}(x) \\ &= r_{(Y\setminus\hat{y})\cup\hat{z},X}, \end{aligned}$$

in which case the pair $(\hat{y}, \hat{z})$ is, according to Definition 15, exchangeable.

As to the converse, suppose the pair $(\hat{y}, \hat{z}) \in Y \times Z$ is exchangeable, i.e., $r_{Y,X} > r_{(Y\setminus\hat{y})\cup\hat{z},X}$. This implies

$$r_{Y,X} > \max_{z\in(Z\setminus\hat{z})\cup\hat{y}} d_{(Y\setminus\hat{y})\cup\hat{z}}(z) \geq \max_{z\in Z\setminus\hat{z}} d_{(Y\setminus\hat{y})\cup\hat{z}}(z) = \max_{z\in Z} d_{(Y\setminus\hat{y})\cup\hat{z}}(z), \tag{4.25}$$

and therefore

$$r_{Y,X} > \max_{z\in Z^*} d_{(Y\setminus\hat{y})\cup\hat{z}}(z) \geq \max_{z\in Z^*} d_{Y\cup\hat{z}}(z),$$

which shows that in this case statement (a) holds. Finally, note that (4.25) immediately implies the statements (b) and (c), which completes our proof. □

Corollary 5. *Let $\hat{z} \in Z$ satisfy condition (a) of Theorem 23. Moreover, let $\hat{y} \in Y$ satisfy $r(\hat{y}) < r_{Y,X}$. Then, the pair $(\hat{y}, \hat{z}) \in Y \times Z$ is exchangeable.*

Proof. Recall the expression $r(\hat{y}) = r_{Y\setminus\hat{y},X\cap V_Y(\hat{y})}$ in (4.16) for the local covering radius of $\hat{y}$, which yields

$$\begin{aligned} r(\hat{y}) &= \max\left(\max_{z\in Z\cap V_Y(\hat{y})} d_{Y\setminus\hat{y}}(z), d_{Y\setminus\hat{y}}(\hat{y})\right) \\ &\geq \max\left(\max_{z\in Z\cap V_Y(\hat{y})} d_{(Y\setminus\hat{y})\cup\hat{z}}(z), d_{(Y\setminus\hat{y})\cup\hat{z}}(\hat{y})\right). \end{aligned}$$

Therefore, the assumption $r_{Y,X} > r(\hat{y})$ directly implies that the pair $(\hat{y}, \hat{z})$ satisfies the conditions (b) and (c) in Theorem 23. In combination with the other assumption on $\hat{z}$, all of the three conditions (a), (b), and (c) in Theorem 23 are satisfied by $(\hat{y}, \hat{z})$. Therefore, the point pair $(\hat{y}, \hat{z})$ is exchangeable. □

In many situations, the set Z^* contains merely one point z^*. In this case, the point $z^* \in Z^* \subset Z$ is potentially a good candidate for an exchange, since it satisfies the condition (a) in Theorem 23. This observation yields, by using the criterion in Corollary 5, the following sufficient condition for an exchangeable pair.

Corollary 6. *Let $z^* \in Z$ satisfy $d_Y(z^*) > d_Y(z)$ for all $z \in Z \setminus z^*$. Then, for $\hat{y} \in Y$, the pair $(\hat{y}, z^*) \in Y \times Z$ is exchangeable, if they satisfy*

$$d_Y(z^*) > r(\hat{y}). \tag{4.26}$$

Proof. Note that the first assumption on z^* implies

$$r_{Y,X} = d_Y(z^*) > d_Y(z) \geq d_{Y \cup z^*}(z), \qquad \text{for all } z \in Z.$$

Hence, the point z^* satisfies the condition (a) of Theorem 23. Moreover, by the other assumption in (4.26), we obtain $r_{Y,X} = d_Y(z^*) > r(\hat{y})$. But in this case, the point pair $(\hat{y}, z^*)$ is, due to Corollary 5, exchangeable. □

4.4.3 Implementation of the Exchange Algorithm

In our implementation of the exchange Algorithm 17 we have, for the sake of simplicity, merely used the sufficient criterion of Corollary 6 for locating exchangeable point pairs. This can be done very efficiently as follows.

Recall the discussion concerning the efficient implementation of greedy thinning in Subsection 4.3.5. As explained in Subsection 4.3.5, greedy thinning works with a heap for maintaining removable points. In this heap, which we call the `Y-heap`, the significance of a node $y \in Y$ is given by the value $r(y)$ of its current local covering radius. We use this heap also for the exchange algorithm.

Moreover, during the performance of the exchange algorithm we use another heap, called the `Z-heap`, where the points of the current set $Z = X \setminus Y$ are stored. The priority of a node $z \in Z$ in the `Z-heap` is given by its distance $d_Y(z)$ to the set Y. The nodes in the `Z-heap` are ordered such that the significance of a node is *greater* than the significances of its two children. Hence, the root of the `Z-heap` contains a point z^* from the set Z^*, so that $d_Y(z^*) = r_{Y,X}$.

We remark that the `Z-heap` may either be built immediately before the performance of the exchange algorithm, or it may be maintained during the performance of the greedy thinning algorithm. In either case, building the `Z-heap` costs at most $\mathcal{O}(N \log N)$ operations. We can explain this as follows. First note that the abovementioned attachments of the points in $Z = X \setminus Y$ to corresponding Voronoi tiles (see Subsection 4.3.5) can be used in order to facilitate this. Indeed, by these attachments the significance $d_Y(z)$ of any $z \in Z \cap V_Y(y)$ is already given by the Euclidean distance between z and $y \in Y$. Now since the number $|Z|$ of points in Z is at most N, and each insertion into the `Z-heap` costs at most $\mathcal{O}(\log N)$ operations, this altogether shows that we require at most $\mathcal{O}(N \log N)$ operations for building the initial `Z-heap`.

Now let us return to the performance of the steps **(2a)** and **(2b)** of the exchange Algorithm 17. In order to locate an exchangeable pair in **(2a)**, we compare the significance $r(y^*)$ of the point y^* (the point in the root of the `Y-heap`) with the significance $d_Y(z^*)$ of z^* (the point in the root of the `Z-heap`). If $r(y^*) < d_Y(z^*)$ and $Z^* = \{z^*\}$, then the pair $(y^*, z^*) \in Y \times Z$ is, due to Corollary 6, exchangeable. Step **(2b)** of the exchange Algorithm 17 is then accomplished as follows.

(E1) Remove y^* from Y by applying greedy thinning on Y. To this end, perform the steps **(T1)**-**(T4)**, described in Subsection 4.3.5.
(E2) Pop the root z^* from the `Z-heap` and update the `Z-heap`.
(E3) Add the point z^* to the Voronoi diagram of the set Y [1] in order to obtain the Voronoi diagram of the set $Y \cup z^*$.
(E4) Update the local covering radii of those points in Y, whose Voronoi tiles were modified by the insertion of z^* in step **(E3)**. Update the positions of these points in the `Y-heap`.
(E5) Update the significances $d_Y(z)$ of those points in Z, whose surrounding Voronoi tile was deleted by the removal of y^* in step **(T2)** or by the insertion of z^* in step **(E3)**. Reattach these points to new Voronoi tiles, and update their positions in the `Z-heap`.
(E6) Let $Y = Y \cup z^*$ and so $Z = Z \setminus z^*$.
(E7) Compute the local covering radius $r(z^*)$ of z^*, and insert z^* into the `Y-heap`.
(E8) Compute the significance $d_Y(y^*)$ of y^*, and insert y^* into the `Z-heap`.

Now let us turn to the computational costs required for *one* exchange step of the exchange Algorithm 17. As explained above, step **(2a)** requires only $\mathcal{O}(1)$ operations, when working with the two heaps, `Y-heap` and `Z-heap`. The performance of one step **(2b)**, as described by the above instructions **(E1)**-**(E8)**, can be done in at most $\mathcal{O}(\log N)$ operations, provided that each Voronoi tile contains $\mathcal{O}(1)$ points from Z. We tacitly act on this reasonable assumption from now. In this case, the required updates of the local covering radii in steps **(E1)**, **(E4)**, and **(E7)** cost only $\mathcal{O}(1)$ time. Likewise, the updates of the significances in steps **(E5)** and **(E8)** cost $\mathcal{O}(1)$ time. Finally, each update in either of the two heaps in steps **(E1)**,**(E2)**,**(E4)**,**(E5)**,**(E7)**, and **(E8)** costs at most $\mathcal{O}(\log N)$ time.

Theorem 24. *One exchange step of the exchange Algorithm 17, by performing the instructions* **(E1)**-**(E8)**, *requires at most* $\mathcal{O}(\log N)$ *operations.* □

We finally remark that we have no (non-trivial) upper bound on the number n_E of exchange steps (required in the exchange Algorithm 17). But in all of our numerical experiments we observed that n_E is always much smaller than the size of the input point set X, i.e., $n_E \ll N = |X|$. We summarize the above results concerning the computational costs of scattered data filtering by combining the Theorems 22 and 24.

[1] Note that at this stage y^* has already been removed from Y by step **(T3)**.

Theorem 25. *For any finite point set X of size $N = |X|$, and $1 \leq n < N$, the construction of the subset $X_{N-n} = E \circ T_*^n(X)$ by n steps of the greedy thinning Algorithm 16 followed by n_E steps of the exchange Algorithm 17 requires at most $\mathcal{O}(N \log N) + \mathcal{O}(n_E \log N)$ operations.* □

4.4.4 Numerical Examples

We have implemented the proposed scattered data filtering scheme in two dimensions, $d = 2$. For the purpose of locating exchangeable point pairs, in step **(2a)** of Algorithm 17, we decided to merely work with the sufficient criterion in Corollary 6, as explained in the previous section. Moreover, our implementation does not remove extremal points from X. This is desirable insofar as this serves to guarantee that the convex hull $[Y]$ of any current subset $Y \subset X$ coincides with the convex hull $[X]$ of the input point set X.

Initially, on given input set X, the significance $\sigma(x)$ in (4.10) is computed for every point $x \in X$. Then, the occurring significances (but not the points!) are sorted in increasing order, so that we obtain the sequence

$$\sigma_1 \leq \sigma_2 \leq \ldots \leq \sigma_N,$$

which is required for recording the quality indices $\alpha_{X_{N-n},X} = r_{X_{N-n},X}/\sigma_n$, where $X_{N-n} = E \circ T_*^n(X)$ or $X_{N-n} = T_*^n(X)$, at run time. Note that this preprocess costs only at most $\mathcal{O}(N \log N)$ operations [33], where $N = |X|$ is the size of X.

The filtering scheme was applied on two different types of scattered data,

- *clustered data* from terrain modelling (Figure 4.2 **(a)**);
- *track data* from marine seismic data analysis (Figure 4.5 **(a)**).

The numerical results on these two examples are discussed in the following of this subsection, one after the other.

Terrain Data. Figure 4.2 **(a)** shows a scattered data sample of a terrain around `Gjøvik`, Norway, comprising $N = 7928$ data points. Note that the sampling density is subject to strong variation. In fact, the data is rather sparse in flat regions of the terrain, whereas a higher sampling rate around steep gradients of the terrain's surface leads to clusters.

For the purpose of graphical illustration, Figure 4.2 shows also the three different subsets **(b)** $F_{2000}(X)$, **(c)** $F_{4000}(X)$, and **(d)** $F_{6000}(X)$, which were generated by using the proposed filtering scheme. The resulting covering radii and the quality indices of $T_*^n(X)$ and $F_n(X)$, $n = 2000, 4000, 6000$, are shown in Table 4.1. Moreover, Table 4.1 shows the CPU seconds $u(T)$ which were required for computing the subsets $T_*^n(X)$ from X by greedy thinning, and the CPU seconds $u(E)$ for the postprocessing exchange of point pairs. Therefore, the sum $u(F) = u(T) + u(E)$ of these values are the total costs, in terms of CPU seconds, for computing the subsets $X_{N-n} = F_n(X)$ from X. The numbers n_E of exchange steps are also shown in Table 4.1.

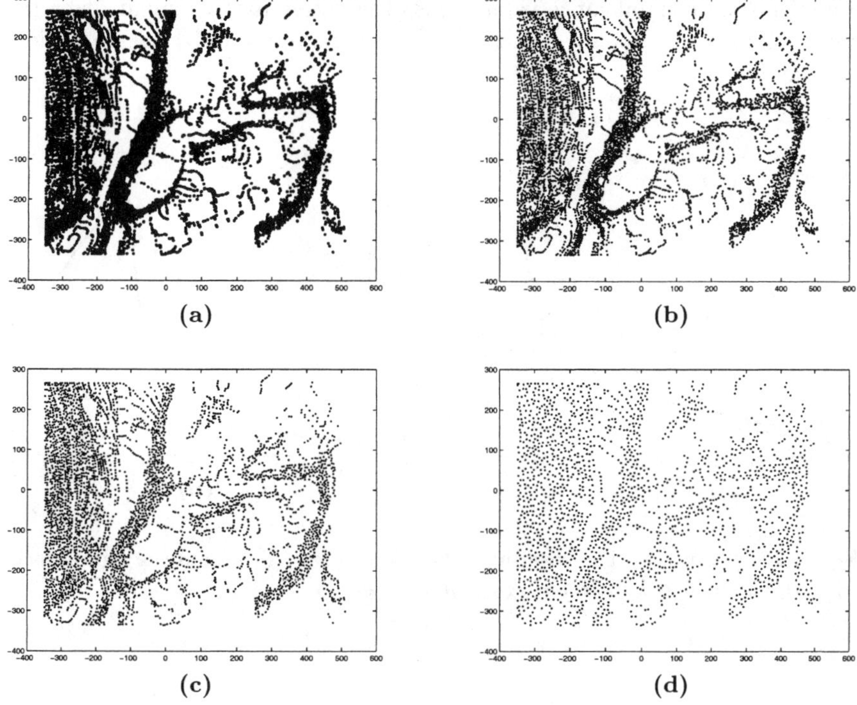

Fig. 4.2. **Gjøvik.** **(a)** The input data set X comprising 7928 points, and the subsets **(b)** X_{2000} of size 5928, **(c)** X_{4000} of size 3928, and **(d)** X_{6000} of size 1928, generated by scattered data filtering.

Table 4.1. Scattered data filtering on **Gjøvik**

n	$r_{T_*^n(X),X}$	$r_{F_n(X),X}$	$\alpha_{T_*^n(X),X}$	$\alpha_{F_n(X),X}$	$u(T)$	$u(E)$	n_E
2000	3.0321	2.9744	1.3972	1.3706	1.86	0.21	71
4000	5.2241	4.6643	1.6462	1.4698	2.84	1.18	409
6000	23.9569	7.9306	5.4281	1.7969	3.74	0.81	381

For further illustration, we have recordered the results in Table 4.1 for *all* possible n. The following Figures 4.3 and 4.4 reflect the results of the entire numerical experiment. The graphs of the resulting covering radii $r_{T_*^n(X),X}, r_{F_n(X),X}$ and the quality indices $\alpha_{T_*^n(X),X}, \alpha_{F_n(X),X}$, $100 \leq n \leq 7391$, are displayed in Figure 4.3. Figure 4.3 (a) shows also the graph of the initial significances σ_n. Recall that $\sigma_n \leq r_n^*$ by Theorem 18, i.e., the value σ_n is a lower bound for the optimal value r_n^*.

We remark that for large values of n the deviation between σ_n and the optimal value r_n^* is typically very large. For $n = N - 1$, for instance, we find $r_{N-1}^* = 516.264$ for the optimal covering radius, but $\sigma_{N-1} = 22.581$ for the penultimate significance value. This observation partly explains why the quality indices of $\alpha_{T_*^n(X),X}$ and $\alpha_{F_n(X),X}$ in Figure 4.3 (**b**) are so rapidly growing for large n.

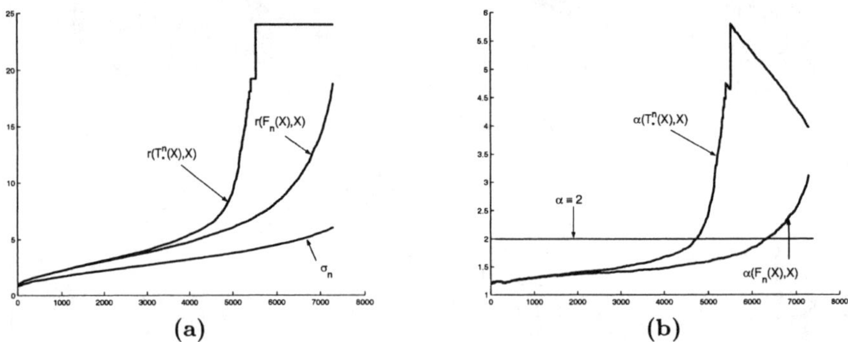

Fig. 4.3. `Gjøvik`. (**a**) The covering radii $r_{T_*^n(X),X}, r_{F_n(X),X}$, and the significances σ_n. (**b**) The quality indices $\alpha_{T_*^n(X),X}$ and $\alpha_{F_n(X),X}$.

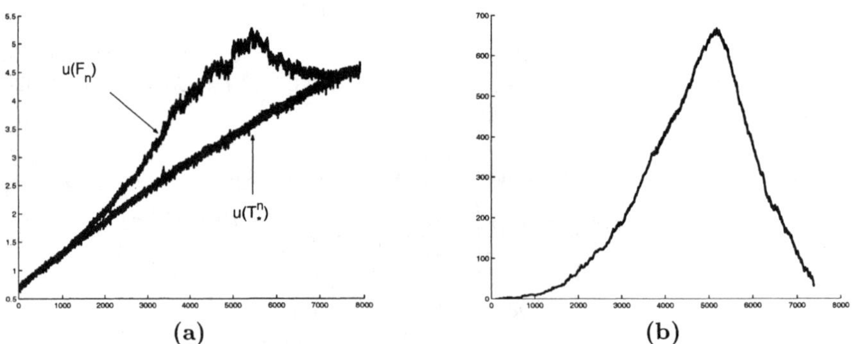

Fig. 4.4. `Gjøvik`. (**a**) CPU seconds $u(F_n)$ required for computing $F_n(X)$, and $u(T_*^n)$ for computing $T_*^n(X)$; (**b**) number of exchange steps.

Nevertheless, for $n \leq 6435$, we found $\alpha(F_n(X), X) < 2$ and moreover, $\alpha(F_n(X), X) < \alpha_2 = \sqrt{2+\sqrt{3}} \approx 1.9319$ for $n \leq 6327$. We mention the latter because for the special case of the Euclidean norm, the best possible constant in (4.7) is $\alpha = \alpha_2$. In other words, there is for $\alpha < \alpha_2$ no α-approximation algorithm for the k-center problem, when using the Euclidean norm, unless P=NP. This result is due to Feder and Greene [67] (see also [164, Section 4]).

In conclusion, the numerical results reflected by Figure 4.3 illustrate the good performance of the proposed filtering scheme, especially in comparison with possible α-approximation algorithms for the k-center-problem. The required seconds of CPU time and the number of exchange steps for computing the sets $X_{N-n} = F_n(X)$ from $T_*^n(X)$ are displayed Figures 4.4 **(a)** and **(b)**. Not surprisingly, we found that the CPU seconds $u(E)$ for the exchange are roughly proportional to the number n_E of exchange steps.

Track Data. In our second numerical experiment, we considered using one example from marine seismic data analysis. In this case, the spatial distribution of the sampled data is organized along *tracks*, since these data are acquired from ships. Figure 4.5 **(a)** shows such a seismic data set which was taken in a region of the North Sea. This data set, here referred to as `NorthSea`, comprises $N = 9758$ data points.

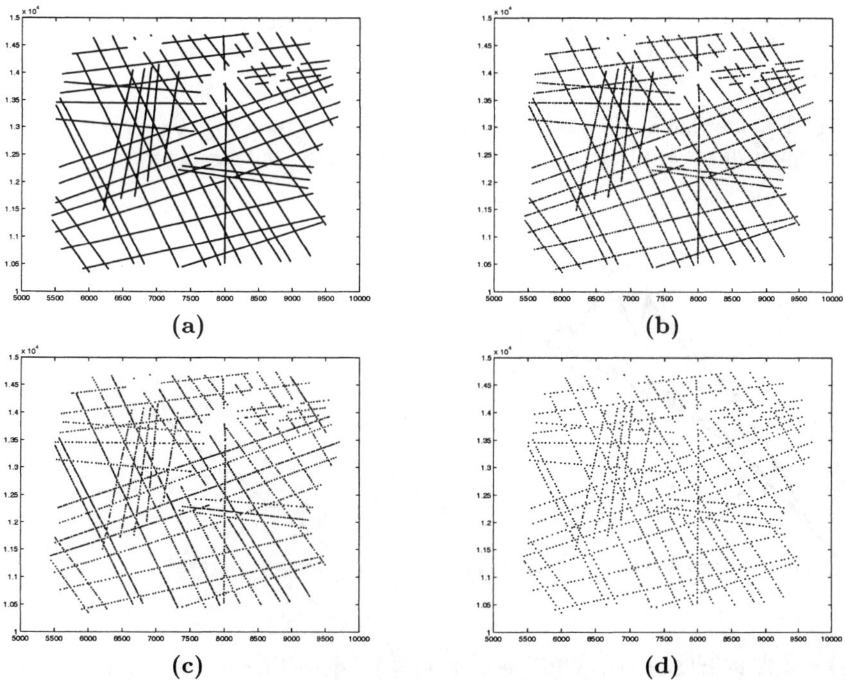

Fig. 4.5. `NorthSea`. **(a)** The input data set comprising 9758 points, and the subsets **(b)** X_{5765} of size 3993, **(c)** X_{6908} of size 2850, and **(d)** X_{8112} of size 1646, generated by scattered data filtering.

We have recorded the covering radii, $r_{T_*^n(X),X}$ and $r_{F_n(X),X}$, and the quality indices, $\alpha_{T_*^n(X),X}$ and $\alpha_{F_n(X),X}$, for all possible n. Figure 4.6 **(a)** displays the graphs of $r_{T_*^n(X),X}$ and $r_{F_n(X),X}$ along with that of the significances σ_n, whereas the graphs of $\alpha_{T_*^n(X),X}$ and $\alpha_{F_n(X),X}$, $1 \leq n \leq 8300$, are shown in Figure 4.6 (**b**). Moreover, we have also recorded the elapsed CPU time required for computing $F_n(X)$ and $T_*^n(X)$, see Figure 4.7 (**a**), as well as the number n_E of exchange steps, which are required for computing $F_n(X)$ from $T_*^n(X)$, see Figure 4.7 (**b**).

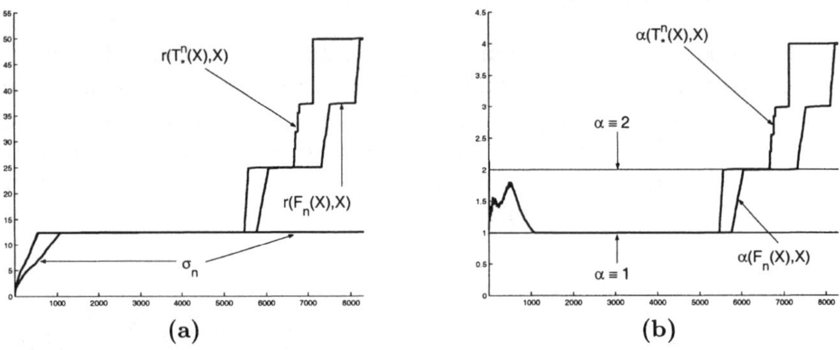

Fig. 4.6. `NorthSea`. The graphs of (**a**) the covering radii $r_{T_*^n(X),X}, r_{F_n(X),X}$ and the significances σ_n; (**b**) the quality indices $\alpha_{T_*^n(X),X}$ and $\alpha_{F_n(X),X}$.

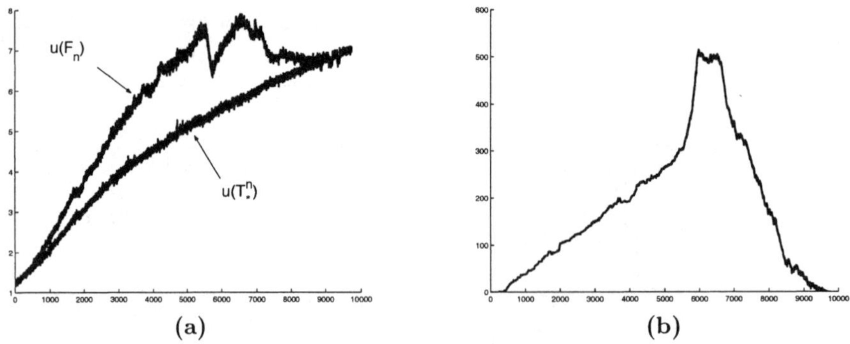

Fig. 4.7. `NorthSea`. (**a**) CPU seconds $u(F_n)$ required for computing $F_n(X)$, and $u(T_*^n)$ for computing $T_*^n(X)$; (**b**) number of exchange steps.

We remark that both greedy thinning and the proposed scattered data filtering scheme perform very well on this data set. This is confirmed by the numerical results concerning the behaviour of the quality indices $\alpha_{T_*^n(X),X}$ and $\alpha_{F_n(X),X}$, see Figure 4.6 (**b**). Indeed, the values $\alpha_{T_*^n(X),X}$ and $\alpha_{F_n(X),X}$ are very close to the best possible value $\alpha \equiv 1$ in the range $1083 \leq n \leq 5472$, where we find

$$1.00155 \le \alpha_{T_*^n(X),X} \le 1.00279, \qquad \text{for all } 1083 \le n \le 5472.$$

The quality index $\alpha_{F_n(X),X}$ continues to be very close to $\alpha \equiv 1$ beyond $n = 5472$, where we find

$$1.00135 \le \alpha_{F_n(X),X} \le 1.00272, \qquad \text{for all } 1083 \le n \le 5765.$$

Moreover, we have $\alpha_{F_n(X),X} < \alpha_2 = \sqrt{2+\sqrt{3}}$ for every $n \le 6032$, and $\alpha_{F_n(X),X} < 2$ for every $n \le 6908$.

The subsets $F_{5765}(X)$ and $F_{6908}(X)$ are shown in Figures 4.5 **(b)** and **(c)**, along with the subset $F_{8112}(X)$, which is displayed in Figure 4.5 **(d)**.

Finally, let us spend a few remarks concerning the results in Figure 4.6.

Firstly, note from Figure 4.6 **(a)** that the significance values σ_n are almost constant for $n \ge 1083$, where we find $12.3987 = \sigma_{1083} \le \sigma_n \le \sigma_N = 12.5005$ for all $1083 \le n \le N$. This is due to the (almost) constant sampling rate of the data acquisition along the track lines. In fact, the smaller significances σ_n, for $n \le 1082$, are attained at sample points near intersections between different track lines.

Secondly, observe from Figure 4.6 **(a)** the step-like behaviour of the covering radii $r_{T_*^n(X),X}$ and $r_{F_n(X),X}$. For the purpose of explaining the jumps in the graph of $r_{T_*^n(X),X}$, let us for the moment assume that the data contains only *one* track line, with a constant sampling rate. In this case, the data points are *uniformly distributed* along one straight line, so that our discussion boils down to greedy thinning on univariate data. But greedy thinning on (uniformly distributed) univariate data is already well-understood [54]. In this case, greedy thinning generates equidistributed subsets of points. To this end, in the beginning the algorithm prefers to remove *intermediate* points, each of whose left and right neighbour have not been removed by the algorithm, yet. Note that the covering radius is then constant. But the point removal leads, after sufficiently many steps, to a situation where the algorithm must remove a point, say y^*, in its next step, whose left and right neighbour have already been removed in previous steps. Now by the removal of y^*, the resulting covering radius will be doubled, which leads to the first jump in the graph of the covering radii. By recursion, the covering radius is kept constant for a while, before the next jump occurs at one later removal, and so on.

Now let us return to the situation of the data set `NorthSea`, which incorporates several track lines. Note that the interferences between the different track lines are rather small. In this case, the recursive point removal by greedy thinning on the separate track lines can widely be done simultaneously. This in turn explains the jumps in the graph of the covering radii $r_{T_*^n(X),X}$ by following along the lines of the above arguments for the univariate case. Note that the postprocessing exchange algorithm can only delay, but not avoid, the jumps of the resulting covering radii of $r_{F_n(X),X}$. This also explains the step-like behaviour of the graph $r_{F_n(X),X}$ in Figure 4.6 **(a)**.

Thirdly, given the almost constant significances σ_n and the jumps in the graphs of $r_{T_*^n(X),X}$ and $r_{F_n(X),X}$, the resulting quality indices $\alpha_{T_*^n(X),X}$ and $\alpha_{F_n(X),X}$ are clearly also subject to jumps by definition, see Figure 4.6 **(b)**. Moreover, we remark that for large n, the differences between the significances σ_n and the optimal covering radii $r_n^*(X)$ are very large, see Figure 4.6 **(a)**. For $n = N-1$, for instance, we find $r_{N-1}^* = 2652.46$ for the optimal covering radius, but $\sigma_{N-1} = 12.5004$. In this case, albeit the adaptive bound in (4.14) is no longer a useful criterion for the subset selection (see the corresponding discussion immediately after Corollary 3), the proposed filtering scheme continues to generate subsets, whose sample points are uniformly distributed along the track lines. One example is given by the subset $F_{8112}(X)$ in Figure 4.5 **(d)**, whose quality index is $\alpha_{F_{8112}(X),X} = 3.0012$.

4.5 Adaptive Thinning

Adaptive thinning is concerned with the approximation of a bivariate function from scattered data by piecewise linear functions over triangulated subsets. To this end, the adaptive thinning process generates subsets of *most significant* points, such that the corresponding hierarchy of piecewise linear interpolants over triangulations of these subsets approximate progressively the function values sampled at the original scattered points. Moreover, the resulting approximation errors are required to be small relative to the number of points in the subsets, so that the error of approximation gradually increases with the reduction in the number of points in the subsets.

This is accomplished by using Algorithm 16, and by working with various specific data-dependent criteria for the point removal in step **(2a)** of Algorithm 16. For the purpose of motivating these adaptive removal criteria, let $X \subset \mathbb{R}^2$ be a finite planar point set and let $f : \mathbb{R}^2 \to \mathbb{R}$ denote an unknown function, whose function values $f\big|_X$, taken at the points in X, are given.

The aim of adaptive thinning is the construction of subsets $Y \subset X$, such that each piecewise linear interpolant $L(f, \mathcal{T}_Y)$ (over a suitable triangulation $\mathcal{T}_Y$ of Y), satisfying

$$L(f, \mathcal{T}_Y)(y) = f(y), \qquad \text{for all } y \in Y,$$

is *close* to the given function values $f\big|_X$ in $\mathbb{R}^N$.

In order to establish this, we require that, for some norm $\|\cdot\|$ on $\mathbb{R}^N$, the *approximation error*

$$\eta(Y, X) = \|L(f, \mathcal{T}_Y)\big|_X - f\big|_X\| \tag{4.27}$$

is small. Note that $\eta(Y, X)$ depends also on both the input values $f\big|_X$ and on the selected triangulation method, but for notational simplicity we omit this. We remark that we mostly work with the Delaunay triangulation $\mathcal{D}_Y$

of Y. In situations of convex data, however, we also use the data-dependent convex triangulation. This is done in Subsection 4.5.2.

Further note that the expression $\eta(Y,X)$ in (4.27) makes only sense, if the convex hull $[Y]$ of $Y \subset X$ coincides with the convex hull $[X]$ of X. In this case, the linear spline interpolant $L(f,\mathcal{T}_Y)$ is well-defined on $[X]$. We ensure this property, $[Y]=[X]$, by not removing *extremal* points from X.

As to the selection of a *suitable* norm $\|\cdot\|$ in (4.27), this essentially should depend on the underlying application. For the choice of the discrete ℓ_∞-norm, for instance, the approximation error $\eta(Y,X)$ becomes

$$\eta_\infty(Y,X) = \max_{x\in X} |L(f,\mathcal{T}_Y)(x) - f(x)|, \tag{4.28}$$

whereas for the discrete ℓ_2-norm, we obtain

$$\eta_2(Y,X) = \sqrt{\sum_{x\in X} |L(f,\mathcal{T}_Y)(x) - f(x)|^2}. \tag{4.29}$$

In either case, the most natural greedy removal criterion, for step **(2a)** of the thinning Algorithm 16, is the following one.

Definition 16. (Removal Criterion AT)
For $Y \subset X$, a point $y^* \in Y$ is said to be `removable` *from Y, iff it satisfies*

$$\eta(Y \setminus y^*, X) = \min_{y\in Y} \eta(Y \setminus y, X).$$

We refer to the adaptive thinning algorithm, resulting from this particular removal criterion, in step **(2a)** of Algorithm 16, as **AT**.

Let us make a few remarks on the idea of this particular definition of a removable point. When using the above removal criterion **AT**, we assign to each current point $y \in Y$ an *anticipated error*

$$e(y) = \eta(Y \setminus y, X), \tag{4.30}$$

which is actually incurred by the removal of y. Similar to the case of *non-adaptive* thinning, we interpret the value $e(y)$ as the significance of the point y in the current subset Y. A point y^* whose removal gives the least anticipated error $e(y^*)$ is considered as *least significant* in the current situation, and so it is removed from Y.

Unlike in non-adaptive greedy thinning, whose removal criterion is given by the local covering radius $r(y)$ in (4.16), the adaptive thinning criterion, given by the anticipated error $e(y)$ in (4.30), depends on both the given function values $f\big|_X$ and the selected triangulation method $\mathcal{T}_Y$.

In the following of this chapter, we propose different removal criteria, which are adapted to terrain modelling and to image compression. Let us briefly explain the differences between these two applications. In terrain modelling, it is of primary importance to keep the *maximal* deviation $\eta_\infty(Y,X)$

between the linear spline interpolant $L(f, \mathcal{D}_Y)$ and the given point samples $f|_X$ as small as possible. This is in contrast to applications in image compression, where the quality measure relies on the *mean square error*

$$\bar{\eta}_2^2(Y, X) = \eta_2^2(Y, X)/N. \tag{4.31}$$

We develop two classes of customized adaptive thinning criteria. Those for terrain modelling, suggested in the following Subsections 4.5.1 and 4.5.4, work with the error measure η_∞ in (4.28), whereas the anticipated error measures for image compression, in Subsection 4.6.2, rely on the discrete ℓ_2-error η_2 in (4.29). Accordingly, we denote by $\mathbf{AT}_\infty$ the adaptive thinning algorithm **AT** which works with the ℓ_∞-norm, whereas $\mathbf{AT_2}$ is the basic algorithm **AT** for the choice of the ℓ_2-norm.

4.5.1 A First Anticipated Error for Terrain Modelling

In this subsection, we propose one removal criterion, **AT1**, whose anticipated error can, unlike $\mathbf{AT}_\infty$, be computed *locally.* This helps to reduce the required computational costs of the resulting thinning algorithm, in comparison with $\mathbf{AT}_\infty$. Later in Subsection 4.5.4, we suggest two alternative removal criteria, **AT2** and **AT3**, which are also based on local computations. A comparison concerning the performance of the algorithms **AT1**, **AT2** and **AT3**, is given in Subsection 4.5.5.

The removal criterion **AT1** measures the anticipated error of a point y only in its cell $C(y)$,

$$e_1(y) = \eta_\infty(Y \setminus y, X \cap C(y)). \tag{4.32}$$

Recall that the cell $C(y)$ of y is the union of all triangles in $\mathcal{T}_Y$ which contain y as a vertex, see Figure 4.1.

The anticipated error measure $e_1(y)$ leads us to the following definition for a removable point in step **(2a)** of Algorithm 16.

Definition 17. (Removal Criterion AT1)
For $Y \subset X$, a point $y^* \in Y$ is said to be `removable` *from Y, iff it satisfies*

$$e_1(y^*) = \min_{y \in Y} e_1(y).$$

Now note that since

$$\begin{aligned} e(y) &= \max\left(\eta_\infty(Y \setminus y, X \cap C(y)), \eta_\infty(Y \setminus y, X \setminus C(y))\right) \\ &= \max\left(e_1(y), \eta_\infty(Y \setminus y, X \setminus C(y))\right), \end{aligned} \tag{4.33}$$

for any $y \in Y$, we obtain the inequality

$$e_1(y) \le e(y), \qquad \text{for all } y \in Y,$$

which gives us a relation for the removal criteria of $\mathbf{AT}_\infty$ and **AT1**.

But we remark that the adaptive thinning algorithm $\mathbf{AT}_\infty$ is not equivalent to **AT1**. This is confirmed by the following counter-example.

Example 1. ($\mathbf{AT}_\infty \neq \mathbf{AT1}$)
Consider the eight data points, $X = \{x_1, \ldots, x_8\}$ and function values $f\big|_X$ given as follows.

i	1	2	3	4	5	6	7	8
x_i	(1,0)	(2,0)	(3,0)	(4,0)	(5,0)	(6,0)	(7,0)	(1,1)
$f(x_i)$	5	−1	0	−3	0	−1.1	2.5	0

In this case, the extremal points of X are x_1, x_7 and x_8, so that only the five points x_i, $i = 2, \ldots, 6$, can be removed. It is easy to see that both $\mathbf{AT}_\infty$ and **AT1** remove the point $x_3 = (3,0)$ first, and then they remove $x_5 = (5,0)$. In the third step, however, the algorithm **AT1** removes $x_6 = (6,0)$, whereas the algorithm $\mathbf{AT}_\infty$ removes $x_4 = (4,0)$.

Let us close this subsection by making a few remarks, mainly concerning comparisons between non-adaptive thinning algorithm of Section 4.3 and the adaptive thinning algorithm **AT1** of this subsection.

First note the correspondence between the two removal criteria for non-adaptive thinning and adaptive thinning, namely the local covering radius $r(y)$ in (4.16), and the anticipated error $e_1(y)$. Both criteria measure the resulting *error*, incurred by the removal of a point y, in the *local* neighbourhood of y. This helps, in either case, to reduce the computational costs of the resulting thinning algorithm.

In non-adaptive thinning, the link between the minimization of the local covering radius $r(y)$ and the minimization of the resulting covering radius $r_{Y\setminus y,X}$ is established in Theorem 20. The key observation there is the equality $r_{Y\setminus y,X} = \max(r_{Y,X}, r(y))$.

But in adaptive thinning, a corresponding link between the minimization of the local anticipated error $e_1(y)$ of **AT1** and the minimization of the (global) *actual error* $e(y)$ of $\mathbf{AT}_\infty$ in (4.30) is missing. In fact, as shown in the above counter-example, the adaptive thinning algorithms $\mathbf{AT}_\infty$ and **AT1** are not equivalent.

We can, however, prove the equivalence of $\mathbf{AT}_\infty$ and **AT1** for the special case of convex data, when working with convex triangulations. This is the result of Theorem 26 in the following Subsection 4.5.2.

Finally, in case the Delaunay triangulation $\mathcal{D}_Y$ is used, it is yet possible to use the anticipated error $e(y)$ as the criterion for the point removal instead of the anticipated error $e_1(y)$ without reducing the computational complexity, at additional computational overhead though. Details on this are discussed in Subsection 4.5.3.

4.5.2 Adaptive Thinning of Convex Data

In this subsection, we discuss the adaptive thinning algorithm **AT1** for the special case where f is a convex function. Clearly, when adaptive thinning is applied on convex data, a natural choice for the triangulation $\mathcal{T}_Y$ is also a convex one. Therefore, throughout this subsection, we work with convex triangulations. In this case, the resulting adaptive thinning algorithm is identical to **AT1** of the previous subsection, except that we always triangulate with a convex triangulation.

Let us briefly recall some important properties of convex triangulations. It has been established in [35] and [161] that if f is any convex function, there is a triangulation $\mathcal{T}_Y$ of Y, for which the piecewise linear interpolant $L(f, \mathcal{T}_Y)$ is convex. This convex triangulation $\mathcal{T}_Y$ of Y is unique if no four data points $(y, f(y))$, $y \in Y$, lie in a plane, and the interpolant $L(f, \mathcal{T}_Y)$ is unique in any case. As shown in [26], decremental convex triangulation requires only retriangulating the corresponding cell $C(y)$, when removing a vertex y from the convex triangulation. This is similar to the Delaunay triangulation, and so this helps to keep the computational costs, required for the thinning algorithm, small.

It is easy to see that the convex interpolant $L(f, \mathcal{T}_Y)$ lies above f, pointwise,

$$f(x) \leq L(f, \mathcal{T}_Y)(x), \qquad \text{for all } x \in [X]. \tag{4.34}$$

Since the interpolant $L(f, \mathcal{T}_Y)$ itself is a convex function, the above inequality (4.34) directly implies

$$L(f, \mathcal{T}_Y)(x) \leq L(f, \mathcal{T}_{Y \setminus y})(x), \qquad \text{for all } x \in [X], \tag{4.35}$$

for every $y \in Y$. Hence, when removing any node $y \in Y$ from the current triangulation $\mathcal{T}_Y$, the resulting piecewise linear interpolant $L(f, \mathcal{T}_{Y \setminus y})$ lies pointwise above the previous $L(f, \mathcal{T}_Y)$. This is similar to the monotonicity $r_{Y \setminus y, X} \leq r_{Y,X}$ for the covering radii in (4.20).

Now recall the anticipated error $e_1(y)$ of **AT1** in (4.32). The above property (4.35) of convex triangulations implies

$$e_1(y) \geq \eta_\infty(Y, X \cap C(y)), \qquad \text{for all } y \in Y.$$

This in turn, together with (4.33), immediately yields for any $y \in Y$ the identity

$$e(y) = \max\left(e_1(y), \eta_\infty(Y, X)\right). \tag{4.36}$$

Note that (4.36) is corresponding to the equality (4.17), appearing as the key observation in the proof of Theorem 20. Likewise, the above identity (4.36) shows that the adaptive thinning algorithm **AT1**, when applied to convex data, minimizes the actual approximation error $e(y)$ at each removal, and so in this case the two adaptive thinning algorithms $\mathbf{AT}_\infty$ and **AT1** are equivalent.

Theorem 26. **($\mathbf{AT}_\infty$ = AT1)**
Suppose f is convex, and let for any subset $Y \subset X$, $\mathcal{T}_Y$ denote the convex triangulation of Y. Then, any removable point $y^* \in Y$, according to the removal criterion **AT1**, *minimizes the actual approximation error $e(y)$ of* $\mathbf{AT}_\infty$ *among all points in Y, i.e.,*

$$e(y^*) = \min_{y \in Y} e(y).$$

Proof. The above observation (4.36) yields the implication

$$e_1(y) \le e_1(\tilde{y}) \quad \Longrightarrow \quad e(y) \le e(\tilde{y}), \qquad \text{for all } y, \tilde{y} \in Y,$$

and so we have $e(y^*) \le e(y)$ for all $y \in Y$, which completes our proof. □

But we can establish another similarity between non-adaptive thinning of Section 4.3 and the adaptive thinning algorithm **AT1** for the special case of convex data. This similarity is concerning adaptive bounds on the approximation error $\eta_\infty(Y, X)$ during the performance of **AT1**. More precisely, we intend to show that the adaptive bounds on the covering radii, proven in Theorem 18 and Corollary 3 of Subsection 4.3.2, directly carry over to adaptive bounds on the approximation error $\eta_\infty(Y, X)$.

But this requires some notational preparations. For any $x \in X$, we denote by

$$\sigma(x) = \eta_\infty(X \setminus x, X) = |L(f, \mathcal{T}_{X \setminus x}(x) - f(x)|$$

the significance of x in X. We assume, without loss of generality, that the points in X are ordered, such that their significances are in increasing order, i.e.,

$$\sigma(x_1) \le \sigma(x_2) \le \cdots \le \sigma(x_N),$$

and we let $\sigma_n = \sigma(x_n)$, $1 \le n \le N$, for notational brevity. Finally, we denote by

$$\eta_n^* = \min_{\substack{Y \subset X \\ |Y| = N - n}} \eta_\infty(Y, X).$$

the least approximation error, which can be attained by a subset $Y \subset X$ of size $|Y| = N - n$.

The result of the following theorem helps us, during the performance of **AT1**, to control, for any current subset $Y \subset X$ of size $|Y| = N - n$, the *relative* deviation

$$\left| \frac{\eta_\infty(Y, X) - \eta_n^*}{\eta_n^*} \right|$$

between the current approximation error $\eta_\infty(Y, X)$ and the optimal value η_n^*.

Theorem 27. *Suppose f is convex, and let for any subset $Y \subset X$, $\mathcal{T}_Y$ denote the convex triangulation of Y. Then, for any subset $Y \subset X$ of size $|Y| = N - n$ the inequalities*

$$\sigma_n \leq \eta_n^* \leq \eta_\infty(Y, X) \leq \alpha_\infty(Y, X) \cdot \eta_n^* \tag{4.37}$$

hold, where $\alpha_\infty(Y, X) = \eta_\infty(Y, X)/\sigma_n \geq 1$.

The key point in the following proof is the monotonicity of the error $\eta_\infty(Y, X)$ in (4.35), which is similar to the monotonicity $r_{Y \setminus y, X} \leq r_{Y,X}$ for the covering radii in (4.20). This observation allows us to widely follow along the lines of the arguments in the proof of Theorem 18, Subsection 4.3.2.

Proof. Let for any $Y \subset X$ of size $|Y| = N - n$, denote by $Z = X \setminus Y \subset X$ the difference set between X and Y of size $|Z| = n$. By the monotonicity (4.35), we have for any $z \in Z$ the inequality

$$|L(f, \mathcal{T}_{X \setminus Z})(z) - f(z)| \geq |L(f, \mathcal{T}_{X \setminus z})(z) - f(z)| = \sigma(z),$$

which implies

$$\begin{aligned} \eta_\infty(Y, X) &= \eta_\infty(X \setminus Z, X) \\ &= \max_{x \in X} |L(f, \mathcal{T}_{X \setminus Z})(x) - f(x)| \\ &= \max_{z \in Z} |L(f, \mathcal{T}_{X \setminus Z})(z) - f(z)| \\ &\geq \max_{z \in Z} \sigma(z). \end{aligned}$$

Now by our assumption on the ordering of the points in X, and by $|Z| = n$, we have

$$\max_{z \in Z} \sigma(z) \geq \sigma(x_n) = \sigma_n,$$

which altogether shows that the inequality

$$\sigma_n \leq \eta_\infty(Y, X)$$

holds, see the corresponding result of Theorem 18. But this immediately implies the inequalities (4.37), see Corollary 3. □

4.5.3 Implementation of Adaptive Thinning

In this subsection, we discuss various computational aspects concerning the *efficient* implementation of the adaptive thinning algorithms **AT1** and **AT**$_\infty$. Moreover, we analyze the complexity of these algorithms. Let us first discuss the implementation of **AT1**.

To this end, we work with decremental Delaunay triangulations, although the following analysis for **AT1** can also be established for the case of convex triangulations, in combination with convex data. In fact, we mainly rely on that after each removal step, the required update on the topology of the subsequent triangulation is *local.* Recall that for either Delaunay triangulations or convex triangulations, this update is done by the (local) retriangulation of the removed point's cell.

The adaptive thinning algorithm **AT1** requires initially computing the Delaunay triangulation $\mathcal{D}_X$ of the input point set X. As explained at the end of Section 2.2, this can be done in $\mathcal{O}(N \log N)$ operations, where $N = |X|$. In addition, like in non-adaptive thinning, we store the nodes of $\mathcal{D}_X$ in a heap, called `Y-heap`, where the significance value of any node $x \in X$ is given by its anticipated error $e_1(x)$.

The nodes in the `Y-heap` are sorted, such that the significance of any node is *smaller* than the significances of its two children. Hence, due to the heap condition, the root of the `Y-heap` contains a least significant node, and thus a removable point, according to the removal criterion of **AT1**. Building the initial `Y-heap` costs $\mathcal{O}(N \log N)$ operations.

As regards the thinning process itself, we assume that throughout the performance of **AT1**, the number of triangles in any cell $C(y)$ of $\mathcal{D}_Y$, $y \in Y \subset X$, is bounded above by a small constant, i.e., is $\mathcal{O}(1)$. Now suppose that $Y \subset X$ is the current subset of size $|Y| = N - n$, which is computed by **AT1** after n removals. Then, the steps **(2a)**,**(2b)** of the thinning Algorithm 16, combined with the removal criterion of **AT1**, are performed as follows.

(T1) Pop the root y^* from the `Y-heap` and update the `Y-heap`.
(T2) Remove the node y^* from the triangulation $\mathcal{D}_Y$ and compute $\mathcal{D}_{Y \setminus y^*}$ by retriangulating the cell $C(y^*)$.
(T3) Attach each point $z \in X \cap C(y^*)$ which has previously been removed, in particular y^* itself, to a *new* triangle in the retriangulated cell $C(y^*)$, which contains z.
(T4) For each neighbouring vertex y of y^* in $\mathcal{D}_Y$, update its anticipated error $e_1(y)$ and its position in the `Y-heap`.

According to step **(T3)**, each of the points in $Z \cap C(y^*)$, $Z = X \setminus Y$, which was removed in a previous step, and the currently removed point y^* is *reattached* to a new triangle in $\mathcal{D}_{Y \setminus y^*}$. Like in non-adaptive thinning, these (re)attachments facilitate the computation and maintenance of the anticipated errors $e_1(y)$ of the points in Y, that have not been removed yet.

As regards the computational costs, step **(T1)** requires at most $\mathcal{O}(\log N)$ operations, step **(T2)** requires $\mathcal{O}(1)$, and step **(T4)** requires at most $\mathcal{O}(\log N)$ operations. As regards step **(T3)**, we assume that the current number of triangles is proportional to $N - n$ and so, provided that the n removed points in $X \setminus Y$ are uniformly distributed over these triangles, the number of operations in step **(T3)** is of the order of $n/(N - n)$ for the n-th step of **AT1**. Therefore, summing the costs of steps **(T1)** to **(T4)** for all n, we find that the total cost for the performance of the adaptive thinning algorithm **AT1** is $\mathcal{O}(N \log N)$.

Now let us turn to $\mathbf{AT}_\infty$. The algorithm $\mathbf{AT}_\infty$ can, like **AT1**, also be implemented at complexity $\mathcal{O}(N \log N)$. But this requires one additional heap, besides the `Y-heap`. This second heap, called the `T-heap`, consists of the current triangles in $\mathcal{D}_Y$, where for any triangle $T \in \mathcal{D}_Y$, its priority $\sigma(T)$ is given by the approximation error

$$\sigma(T) = \eta_\infty(Y, X \cap T).$$

The triangles in the `T-heap` are ordered such that the root of the `T-heap` points to a triangle T^* whose approximation error is largest, i.e., $\sigma(T^*) = \max_{T \in \mathcal{D}_Y} \sigma(T)$. Now from (4.33), it is easy to see that there must be a removable point, according to the removal criterion $\mathbf{AT}_\infty$, among y^* and the three vertices of the triangle T^*.

As to the complexity of $\mathbf{AT}_\infty$, note that we have to maintain both the `T-heap` and the `Y-heap` during the performance of $\mathbf{AT}_\infty$. Our estimates show that the number of required operations in the modified step **(T4)** is significantly increased. Nevertheless, the adaptive thinning algorithm $\mathbf{AT}_\infty$ still requires only $\mathcal{O}(N \log N)$ operations, but with a much larger constant than **AT1**.

We summarize the discussion of this subsection as follows.

Theorem 28. *The performance of the thinning Algorithm 16, in combination with either the adaptive removal criterion* **AT1** *or with* $\mathbf{AT}_\infty$*, and according to the steps* **(T1)**-**(T4)***, requires at most* $\mathcal{O}(N \log N)$ *operations.* □

4.5.4 Two Alternative Anticipated Errors for Terrain Modelling

In this subsection, we propose two possible simplifications of the adaptive thinning algorithm **AT1**, the adaptive thinning algorithms **AT2** and **AT3**. Both alternative removal criteria are locally computable. While the removal criterion of **AT2** works, like **AT1**, with (local) retriangulations of Delaunay cells, the other alternative, **AT3**, does not require such local retriangulations. In fact, the removal criterion of **AT3** is much simpler, and the adaptive thinning algorithm **AT3** is much faster than **AT2** and **AT1**. This is supported by the numerical examples in the following Subsection 4.5.5, where the performance of the three different adaptive thinning algorithms, **AT1**, **AT2** and **AT3**, is compared.

Let us first explain the adaptive thinning algorithm **AT2**. In order to further reduce the required computational costs of **AT1**, the removal criterion **AT2** depends only on the sample values $f\big|_Y$ of the points in the current subset $Y \subset X$. This is in contrast to both $\mathbf{AT}_\infty$ and **AT1**, each of which depends on points in X that were removed in previous steps.

The anticipated error of **AT2** is, for any $y \in Y$, given by

$$e_2(y) = \eta_\infty(Y \setminus y; Y).$$

Note that this expression can be rewritten as

$$e_2(y) = |L(Y \setminus y; f)(y) - f(y)|.$$

Definition 18. (Removal Criterion AT2)
For $Y \subset X$, a point $y^* \in Y$ is said to be `removable` *from Y, iff it satisfies*

$$e_2(y^*) = \min_{y \in Y} e_2(y).$$

As to the other alternative adaptive thinning algorithm, **AT3**, this one does not only ignore the points which were removed in previous steps. Unlike **AT1** and **AT2**, the algorithm **AT3** does also avoid temporary local retriangulations of cells when updating current anticipated errors.

Instead of this, the anticipated error of **AT3** works with *directional anticipated errors* at the current points, in a certain sample of directions. In order to further explain this particular error measure, let $y \in Y$ denote a node in the Delaunay triangulation $\mathcal{D}_Y$ of the current subset $Y \subset X$. For each neighbouring vertex z of y in $\mathcal{D}_Y$ we consider the unique point p lying at the intersection of the boundary of $C(y)$ and the straight line passing through z and y (other than z itself). Such a point exists, since $C(y)$ is a *star-shaped polygon.*

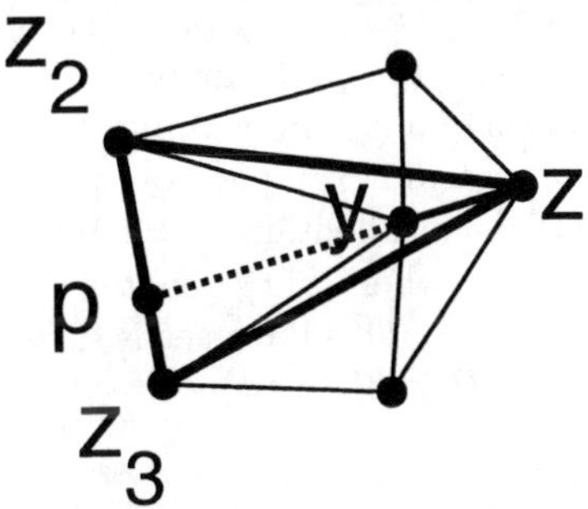

Fig. 4.8. Directional triangle of x.

The point p is either a vertex of the cell's boundary $\partial C(y)$ or a point on one of its sides. In either case, p lies on at least one edge of $\partial C(y)$. Let us denote such an edge by $[z_2, z_3]$; see Figure 4.8. Then the triangle $T_z = [z, z_2, z_3]$, with vertices in $Y \setminus y$, contains y. We call T_z a *directional triangle* of y. We then let

$$e_3^z(y) = |L(f, T_z)(y) - f(y)|$$

be the (unique) directional anticipated error of y in the direction $z - y$, where $L(f, T_z)$ is the linear function which interpolates f at the vertices of T_z. Now, we let

$$e_3(y) = \max_{z \in V_y} e_3^z(y)$$

for the anticipated error of the adaptive thinning algorithm **AT3**, where V_y is the set of all neighbouring vertices of y in $\mathcal{D}_Y$.

Definition 19. (Removal Criterion AT3)
For $Y \subset X$, *a point* $y^* \in Y$ *is said to be* `removable` *from* Y, *iff it satisfies*

$$e_3(y^*) = \min_{y \in Y} e_3(y).$$

We remark that the complexity of the two alternative adaptive thinning algorithms, **AT2** and **AT3**, is $\mathcal{O}(N \log(N))$. This can easily be verified by following along the lines of the arguments in the previous Subsection 4.5.3. But for the sake of brevity, we refrain from doing so here.

4.5.5 Adaptive Thinning of Terrain Data

We have implemented the thinning algorithms **AT1**, **AT2**, and **AT3** together with one earlier version of the non-adaptive thinning algorithm in Section 4.3, called **NAT**, which is developed in our previous paper [74]. Recall that any non-adaptive thinning algorithm, such as **NAT**, ignores the given samples $f|_X$, and so it usually favours evenly distributed subsets $Y \subset X$.

In this section we compare the performance of these four algorithms in terms of both approximation quality and computational cost. To this end, we have considered using one specific example from terrain modelling. The corresponding data set, *Hurrungane*, contains N=23092 data points. Each data point is of the form $(x, f(x))$, where $f(x)$ denotes the terrain's height value sampled at the location $x \in \mathbb{R}^2$. This data set is displayed in Figure 4.9 **(a)** (2D view) and in Figure 4.9 **(b)** (3D view).

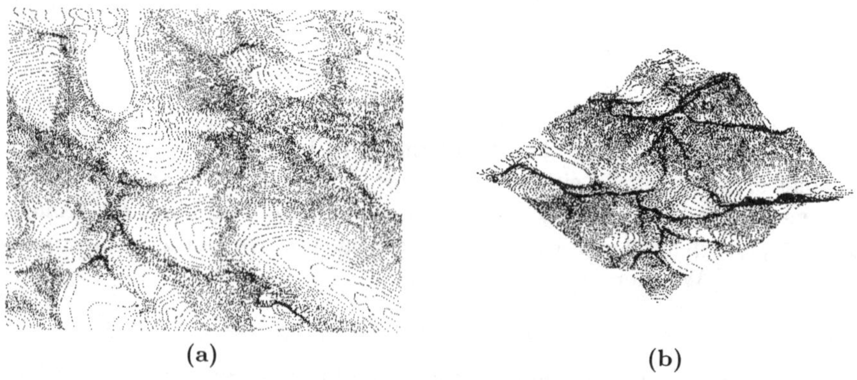

(a) (b)

Fig. 4.9. Hurrungane: 2D view **(a)** and 3D view **(b)**.

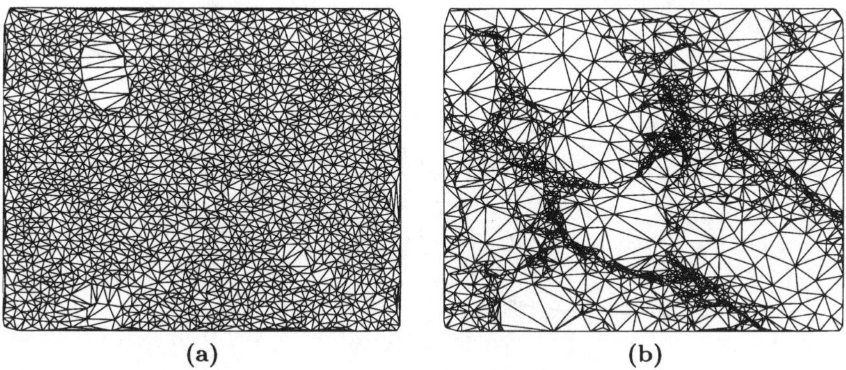

Fig. 4.10. Thinned Hurrungane with 1092 points, 2D view, by using **(a) NAT** and **(b) AT1**.

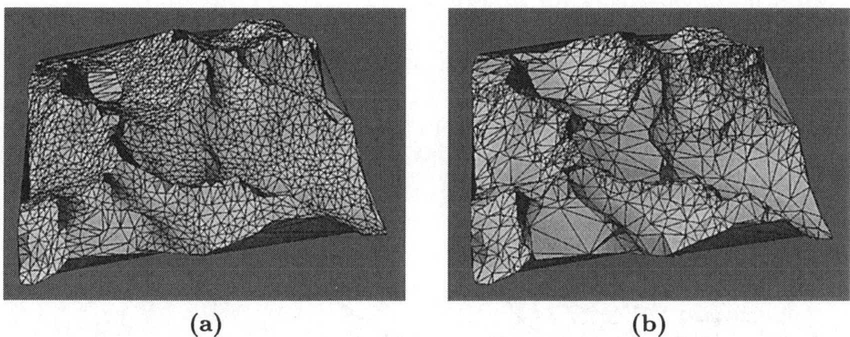

Fig. 4.11. Thinned Hurrungane with 1092 points, 3D view, by using **(a) NAT** and **(b) AT1**.

For all four thinning algorithms, we have recorded both the required seconds of CPU time (without considering the computational costs required for building the initial Delaunay triangulation $\mathcal{D}_X$ and the `Y-heap`) and the sequence of approximation errors $\eta_\infty(Y, X)$ after the removal of $n = 1000, 2000, \ldots, 22000$ points. Not surprisingly, we found that **NAT** is the fastest method but also the worst one in terms of its approximation error. For example, for $n = 22000$ the algorithm **AT1** takes 247.53 seconds of CPU time, whereas **NAT** takes only 11.37 seconds. On the other hand, we obtain in this particular example $\eta_\infty(Y, X) = 278.61$ for **NAT**, but only $\eta_\infty(Y, X) = 30.09$ when using **AT1**. The two corresponding triangulations $\mathcal{D}_Y$ output by **NAT** and **AT1** are displayed in Figure 4.10 **(a)** and **(b)** (2D view), and in Figure 4.11 **(a)** and **(b)** (3D view).

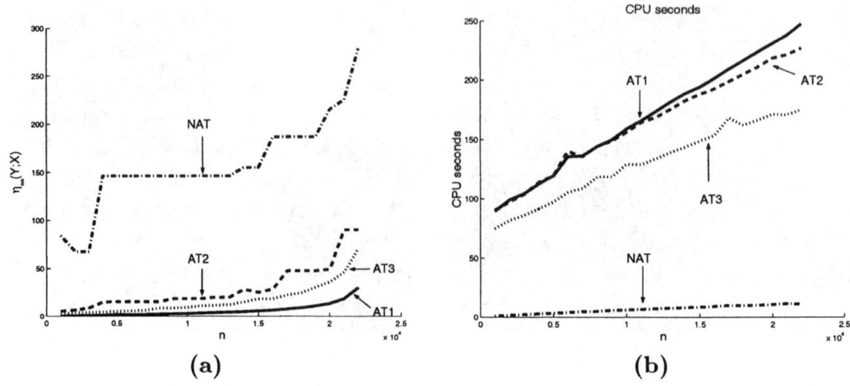

(a) (b)

Fig. 4.12. Hurrungane: comparison between **NAT** (dash-dot line), **AT1** (solid), **AT2** (dashed), and **AT3** (dotted), approximation error **(a)** and seconds of CPU time **(b)**.

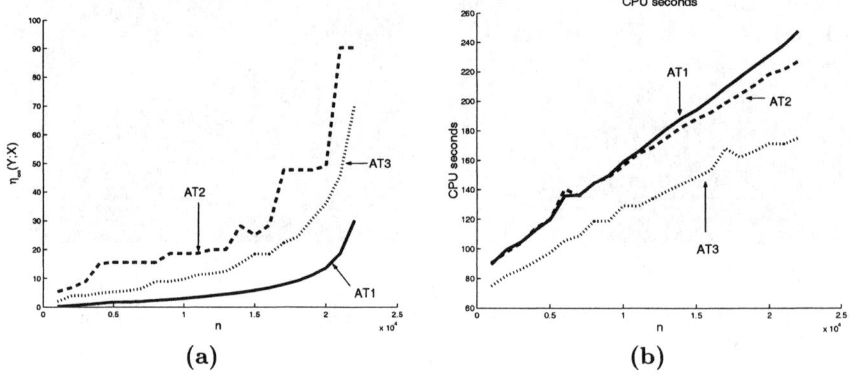

(a) (b)

Fig. 4.13. Hurrungane: comparison between **AT1** (solid line), **AT2** (dashed), and **AT3** (dotted), approximation error **(a)** and seconds of CPU time **(b)**.

In Figure 4.12 **(a)** and in Figure 4.13 **(a)** the approximation error as a function of the number of removed points is plotted for the different thinning algorithms. In Figure 4.12 **(b)** and in Figure 4.13 **(b)** the corresponding seconds of CPU time are displayed.

The graphs show that, with respect to approximation error, the three adaptive thinning algorithms **AT1**, **AT2**, and **AT3** are much better than **NAT**. Among the three adaptive thinning algorithms, **AT1** is the best, followed by **AT3**, and lastly **AT2**. Note that by definition **AT3** can only be inferior to **AT2** after one removal. In the numerical examples **AT3** has continued to be inferior for about 50 removal steps, after which its approximation error is smaller than that of **AT2**.

As to the computational costs for the adaptive thinning algorithms, **AT3** is the fastest, and **AT1** the slowest, see Figure 4.12 **(b)** and Figure 4.13 **(b)** Our conclusion is that **AT1** is our recommended thinning algorithm. But if computational time is a critical issue, **AT3** is a good alternative.

4.6 Adaptive Thinning in Digital Image Compression

4.6.1 Basic Concepts

Digital image compression has recently gained enormous popularity in a variety of applications. This is especially due to the rapid development of multimedia technologies. Digital image compression is concerned with the conversion of digital image data into a *bitstream*. The required information reduction and efficient coding of digital images are essential for fast transmission across an information channel, such as the internet. For the sake of transmission speed, the length of the bitstream message is required to be as short as possible, while maintaining a reasonable quality of the image. For a comprehensive introduction to image compression and coding, we recommend the textbook [167].

Any compression scheme is mainly concerned with the following sequence of tasks.

(1) Data reduction;
(2) Encoding of the reduced data at the sender;
(3) Transmission of the encoded data from the sender to the receiver;
(4) Decoding of the transmitted data at the receiver;
(5) Data reconstruction.

Many of the well-established methods in digital image compression, including JPEG2000 [167] and SPIHT [145], are based on wavelets and related techniques. For a comprehensive introduction to image compression by using wavelets, we refer to the survey [37]. Tensor product wavelets provide, in combination with well-adapted compression methods, high compression rates while maintaining most of the visual features of the image. At low bit rates, however, such methods typically fail to capture certain characteristic features of the image, such as sharp edges. This is mainly due to high oscillations around discontinuities which leads to undesired visual artifacts. Therefore, current research is focussing on the modelling of sharp edges. Recent work on this includes nonlinear decomposition techniques [29, 30, 112].

This section proposes an alternative concept for *lossy* compression of digital image data, by using adaptive thinning methods in combination with *scattered data coding*. Adaptive thinning, when applied on a digital image, leads to a reduction of the original data, step **(1)**. This is accomplished by recursively deleting pixels of the image, so that adaptive thinning returns a *scattered* subset of *most significant* pixels. These most significant pixels are then used in step **(5)** for the reconstruction of the given image. We remark that adaptive thinning provides a class of *data-dependent* filtering operators. In contrast to data-independent filtering methods, the data-dependent approach taken here copes very well with the visual perception of the image. This is supported by the numerical results in Subsection 4.6.6. Further details concerning the application of adaptive thinning on digital images, especially the relevant background for the steps **(1)** and **(5)**, are explained in the following Subection 4.6.3.

The performance of the intermediate steps **(2)**-**(4)** requires an efficient scheme for scattered data coding. This is in order to *encode* and *decode* the most significant pixels. To this end, we have designed a customized scattered data coding scheme, which serves to convert the most significant pixels into a bitstream message, to be transmitted in step **(3)**. This coding scheme is subject of the discussion in Subsection 4.6.4.

We remark that the resulting compression algorithm, proposed in this section, incorporates a flexible representation of the *gridded* image data by using a *scattered* set of most significant pixels. This, in combination with the customized scattered data coding scheme, allows us to capture sharp edges and related image features reasonably well, at small coding costs and at small computational costs. The latter is supported by the analysis concerning the complexity of the compression and decompression, which is done in Subection 4.6.5. Finally, the good performance of the proposed compression scheme is confirmed by numerical examples in Section 4.6.6, where we compare our method with the well-established wavelet-based image compression algorithm SPIHT.

4.6.2 Anticipated Errors for Image Compression

In this subsection, we propose two different removal criteria for image compression, called $\mathbf{AT_2}$ and $\mathbf{AT_2^2}$. Each of these two removal criteria works with the square error η_2^2 of the discrete ℓ_2-norm in (4.29). The basic difference between $\mathbf{AT_2}$ and $\mathbf{AT_2^2}$ is that $\mathbf{AT_2}$ is a greedy *one-point* removal scheme, whereas $\mathbf{AT_2^2}$ is a greedy *two-point* removal scheme. In other words, the algorithm $\mathbf{AT_2^2}$ removes, unlike $\mathbf{AT_2}$, a (removable) pair of two current points at each step **(2a)** of the thinning Algorithm 16. In this sense, $\mathbf{AT_2^2}$ not as short-sighted as $\mathbf{AT_2}$, which leads, in all our test cases below, to significant improvements concerning the performance of the resulting compression scheme.

Now let us first motivate the construction of $\mathbf{AT_2}$ and $\mathbf{AT_2^2}$. A well-known quality measure for the evaluation of image compression schemes is the *Peak Signal to Noise Ratio* (PSNR) in (4.41), see the discussion in the following Subsection 4.6.3. It is sufficient for the moment to say that the PSNR is an equivalent reciprocal measure to the *mean square error*

$$\bar{\eta}_2^2(Y;X) = \frac{1}{N} \sum_{x \in X} |L(f,\mathcal{D}_Y)(x) - f(x)|^2. \tag{4.38}$$

The aim of adaptive thinning, when applied to image compression, is to keep the mean square error as small as possible. This is accomplished by using adaptive thinning algorithms, which generate subsets $Y \subset X$ in (4.1), whose square error

$$\eta_2^2(Y;X) = \sum_{x \in X} |L(f,\mathcal{D}_Y)(x) - f(x)|^2$$

is, among all subsets of equal size, small. Therefore, we prefer to work with the discrete ℓ_2-norm η_2 rather than with any other possible norm in (4.27).

Let us first discuss the adaptive thinning algorithm $\mathbf{AT_2}$, whose anticipated error is given by

$$e(y) = \eta_2^2(Y \setminus y; X), \qquad \text{for } y \in Y.$$

By the additivity of η_2^2 and by the observations $X = (X \setminus C(y)) \cup (X \cap C(y))$, and $\eta_2^2(Y \setminus y; X \setminus C(y)) = \eta_2^2(Y; X \setminus C(y))$, for any $y \in Y$, we get

$$\begin{aligned} \eta_2^2(Y \setminus y; X) &= \eta_2^2(Y \setminus y; X \setminus C(y)) + \eta_2^2(Y \setminus y; X \cap C(y)) \\ &= \eta_2^2(Y; X \setminus C(y)) + \eta_2^2(Y \setminus y; X \cap C(y)) \\ &= \eta_2^2(Y;X) + \eta_2^2(Y \setminus y; X \cap C(y)) - \eta_2^2(Y; X \cap C(y)). \end{aligned}$$

Hence, for any $Y \subset X$, the minimization of $\eta_2^2(Y \setminus y; X)$ is equivalent to minimizing the difference

$$e_\delta(y) = \eta_2^2(Y \setminus y; X \cap C(y)) - \eta_2^2(Y; X \cap C(y)), \qquad \text{for } y \in Y,$$

where $C(y)$ is the cell of y in $\mathcal{D}_Y$.

Definition 20. (Removal Criterion $\mathbf{AT_2}$)
For $Y \subset X$, a point $y^* \in Y$ is said to be `removable` *from Y, iff it satisfies*

$$e_\delta(y^*) = \min_{y \in Y} e_\delta(y).$$

Next we propose another removal criterion, $\mathbf{AT_2^2}$, which requires more computational time than $\mathbf{AT_2}$, but is more effective. The algorithm $\mathbf{AT_2^2}$ works with the greedy removal of *two* points at one step of adaptive thinning.

For any point pair $y_1, y_2 \in Y$, the adaptive error measure $\mathbf{AT_2^2}$ is given by

$$e(y_1, y_2) = \eta_2^2(Y \setminus \{y_1, y_2\}; X) - \eta_2^2(Y;X), \qquad \text{for } y_1, y_2 \in Y. \tag{4.39}$$

Note that the removal in $\mathbf{AT_2^2}$ is either done by the removal of the edge $[y_1, y_2]$ from the triangulation $\mathcal{D}_Y$, and the subsequent retriangulation of

$C(y_1) \cup C(y_2)$, provided that the points $y_1, y_2 \in Y$ are connected by an edge in $\mathcal{D}_Y$, or otherwise by two consecutive greedy removals according to the removal criterion $\mathbf{AT_2}$. To see the latter, suppose that the two points $y_1, y_2 \in Y$ are *not* connected by an edge in $\mathcal{D}_Y$. In this case, the two cells $C(y_1)$ and $C(y_2)$ have no common triangle, and therefore

$$\eta_2^2(Y \setminus y_2, X) - \eta_2^2(Y, X) = \eta_2^2(Y \setminus \{y_1, y_2\}, X) - \eta_2^2(Y \setminus y_1, X),$$

which implies

$$\begin{aligned} e(y_1, y_2) &= \eta_2^2(Y \setminus \{y_1, y_2\}, X) - \eta_2^2(Y, X) \\ &= \eta_2^2(Y \setminus \{y_1, y_2\}, X) - \eta_2^2(Y \setminus y_1, X) + \eta_2^2(Y \setminus y_1, X) - \eta_2^2(Y, X) \\ &= \eta_2^2(Y \setminus y_2, X) - \eta_2^2(Y, X) + \eta_2^2(Y \setminus y_1, X) - \eta_2^2(Y, X) \\ &= e_\delta(y_2) + e_\delta(y_1). \end{aligned}$$

This shows that the anticipated error of $\mathbf{AT_2^2}$ can be simplified as

$$e(y_1, y_2) = e_\delta(y_1) + e_\delta(y_2), \tag{4.40}$$

if $y_1, y_2 \in Y$ are *not* connected by an edge in $\mathcal{D}_Y$.

We remark that the simpler representation (4.40) for the error measure of $\mathbf{AT_2^2}$ helps to reduce the computational costs required for the maintenance of the significance values $e(y_1, y_2)$. Indeed, this merely requires maintaining the significances $e_\delta(y)$, $y \in Y$, in a priority queue, in combination with another priority queue for the edges in $\mathcal{D}_Y$, where their significances, given by the error measure $e(\cdot, \cdot)$ in (4.39), are stored.

Definition 21. (Removal Criterion $\mathbf{AT_2^2}$)
For $Y \subset X$, a point pair $y_1^*, y_2^* \in Y$ is said to be `removable` *from Y, iff it satisfies*

$$e(y_1^*, y_2^*) = \min_{y_1, y_2 \in Y} e(y_1, y_2).$$

4.6.3 Adaptive Thinning in Image Reduction and Reconstruction

This section explains how we use adaptive thinning for the data reduction step **(1)** and the subsequent reconstruction in step **(5)**. To this end, we work with the two adaptive thinning algorithms $\mathbf{AT_2}$ and $\mathbf{AT_2^2}$ of the previous subsection. But let us first explain some details concerning the representation of digital images, before we discuss the application of adaptive thinning to digital images.

A digital image is a rectangular grid of *pixels*. Each pixel bears a color value or greyscale *luminance*. For the sake of simplicity, we restrict the following discussion to greyscale images. The image can be viewed as a matrix $F = (f(i,j))_{i,j}$, whose entries $f(i,j)$ are the luminance values at the pixels. The pixel positions (i,j) are pairs of non-negative integers i and j, whose range is often of the form $[0..2^p - 1] \times [0..2^q - 1]$, for some positive integers p, q, where we let $[0..n] = [0, n] \cap \mathbb{Z}$ for any non-negative integer $n \in \mathbb{Z}$. In this case, the size of the matrix F is $2^p \times 2^q$. Likewise, the entries $f(i, j)$ in F are non-negative integers whose range is typically $[0..2^r - 1]$, for some positive integer r. In the examples of the test images below, we work with 256 greyscale luminances in $[0..255]$, so that in this case $r = 8$.

Adaptive thinning, when applied to digital images, recursively deletes pixels using the thinning Algorithm 16, in combination with the adaptive removal criteria $\mathbf{AT_2}$ or $\mathbf{AT_2^2}$ of Section 4.6.2. In other words, the pixel positions form the initial point set X on which the adaptive thinning algorithm is applied. At any step of the algorithm, a *removable pixel* (point) is removed from the image. The output of adaptive thinning is a set $Y \subset X$ of pixels combined with their corresponding luminances F_Y.

However, due to the regular distribution of pixel positions X, the Delaunay triangulations of X, and of its subsets $Y \subset X$ might be non-unique. To avoid this ambiguity, we apply a small perturbation to the pixels X and apply the thinning algorithm to the perturbed pixels.

A well-known quality measure for the evaluation of image compression schemes is the *Peak Signal to Noise Ratio* (PSNR),

$$\text{PSNR} = 10 * \log_{10}\left(\frac{2^r \times 2^r}{\bar{\eta}_2^2(Y; X)}\right), \tag{4.41}$$

which is an equivalent measure to the reciprocal of the mean square error $\bar{\eta}_2^2(Y; X)$ in (4.38).

The PSNR is expressed in dB (decibels). Good image compressions typically have PSNR values of 30 dB or more [167] for the reconstructed image. The popularity of PSNR as a measure of *image distortion* derives partly from the ease with which it may be calculated, and partly from the tractability of linear optimization problems involving squared error metrics. More appropriate measures of *visual distortion* are discussed in [167].

As a postprocess to the thinning, we further minimize the mean square error by *least squares approximation* [14]. More precisely, we compute from the output set Y and the values F the unique *best ℓ_2-approximation* $L^*(F, \mathcal{D}_Y) \in \mathcal{S}(\mathcal{D}_Y)$ satisfying

$$\sum_{(i,j)\in X} |L^*(F, \mathcal{D}_Y)(i, j) - f(i, j)|^2 = \min_{s\in\mathcal{S}(\mathcal{D}_Y)} \sum_{(i,j)\in X} |s(i, j) - f(i, j)|^2. \tag{4.42}$$

Such a unique solution exists since $Y \subset X$. The compressed information to be transferred consists of the output set Y and the corresponding optimized luminances $\{f^*(i, j) = L^*(F, \mathcal{D}_Y)(i, j) \,:\, (i, j) \in Y\}$.

Following along the lines of our papers [41, 42], we apply a uniform quantization to these *optimized* luminances. This yields the quantized symbols $Q(f^*(i, j))$, $(i, j) \in Y$, corresponding to the quantized luminance values $\tilde{f}(i, j) \approx f^*(i, j)$, for all $(i, j) \in Y$. The set $\{(i, j, Q(f^*(i, j))) \,:\, (i, j) \in Y\}$ is coded by using the customized scattered data coding scheme of the following Subsection 4.6.4.

At the receiver, the reconstruction of the image F, step **(5)**, is then accomplished as follows. The *unique* Delaunay triangulation $\mathcal{D}_Y$ of the pixel positions Y is computed at the decoder, using the same perturbation rules applied previously at the encoder. This defines, in combination with the decoded luminance values $\tilde{F}_Y = \{\tilde{f}(i,j) : (i,j) \in Y\}$, the unique linear spline $\tilde{L}(\tilde{F}_Y, \mathcal{D}_Y) \in \mathcal{S}(\mathcal{D}_Y)$ satisfying $\tilde{L}(\tilde{F}_Y, \mathcal{D}_Y)(i,j) = \tilde{f}(i,j)$ for every $(i,j) \in Y$. Finally, the reconstruction of the image is given by the image matrix $\tilde{F} = (\tilde{L}(\tilde{F}_Y, \mathcal{D}_Y)(i,j))_{(i,j)\in X}$.

4.6.4 Scattered Data Coding

In this section, we explain how the construction of the bitstream in step **(2)** is done. This is concerning the coding of the most significant pixels in Y. The bitstream will contain the elements of the set $\{(i,j,Q(f^*(i,j))) : (i,j) \in Y\}$. Since we work with (unique) Delaunay triangulations, no connectivity coding is required. This helps us to keep the bitstream short. Before we explain details on our coding scheme, let us make some remarks.

First of all, note that the *uncompressed* code for these data (without using quantization) would consist of a header containing the dimension of the image matrix F, followed by the corresponding most significant pixel positions $(i,j) \in Y$ and luminance values $f^*(i,j) \in [0..2^r - 1]$. Thus, when sending uncompressed data, the total coding costs (without the costs for the header) are $(2p + r) \times |Y|$ bits, where $|Y|$ is the number of most significant pixels. This naive way of coding is too costly. In order to reduce the coding costs, we take advantage of the statistical data distribution.

We remark that classical wavelet methods usually consider the complete set of coefficients and decide, according to some suitable threshold, whether a value is significant or not. Then, sophisticated techniques take advantage of the dependencies between the locations and the magnitudes of significant coefficients. This is done by clustering the non-significant coefficients (*zerotrees* [145]), or by using *context-based arithmetic coding* [166].

Our coding strategy exhibits some similarities to this. Indeed, the selected pixel positions in Y are classified as most significant, whereas the remaining pixels are regarded as non-significant. On the other hand, the values $f^*(i,j)$ of the most significant pixels are not proportional to their significances. This requires taking into account the local correlations between the most significant pixels $(i,j,f^*(i,j))$, $(i,j) \in Y$.

Moreover, we apply a uniform *quantization* on the luminance values $f^*(i,j)$, yielding quantized symbols $Q(f^*(i,j))$ for all $(i,j) \in Y$, where the quantization step depends on a specific target rate. This reduces the range of the luminance values, from previously $[0..2^r - 1]$ to $[0..2^s - 1]$ for the quantized symbols, where $s < r$. Now the pixels (positions and quantized symbols) can be viewed as a set of tridimensional points $(i,j,k) \in \Omega$, where we let

$$\Omega = [0..2^p - 1] \times [0..2^p - 1] \times [0..2^s - 1]$$

denote the *bounding cell* of the data, containing $m = |Y|$ pixels. For any such data point $(i, j, k) \in \Omega$, we have $k = Q(f^*(i, j))$ and $(i, j) \in Y$. This data representation can also be viewed as a *binary* tridimensional matrix $M = (m_{ijk})_{i,j,k}$, of dimension 2^p-by-2^p-by-2^s, whose entries are given by

$$m_{ijk} = \begin{cases} 1, & \text{if } (i, j) \in Y \text{ and } k = Q(f^*(i, j)), \\ 0, & \text{otherwise.} \end{cases}$$

In what follows, we propose an efficient coding scheme for the *sparse* matrix M. Note that for coding M, it is sufficient to localize the nonzero entries of M, the *fill-ins* of M. To this end, we work with a hierarchical subdivision of cells. Initially, the bounding cell Ω is split into eight subcells. When splitting Ω, the number of pixels in the resulting subcells are progressively coded. Initially, we code $m = |\Omega|$, the total number of pixels in the bounding cell Ω. Since these are at most $2^p \times 2^p$ pixels, the coding of m requires $2p$ bits, yielding the first $2p$ bits in the bitstream. Then, the splitting of Ω is done in three stages as follows, see Figure 4.14.

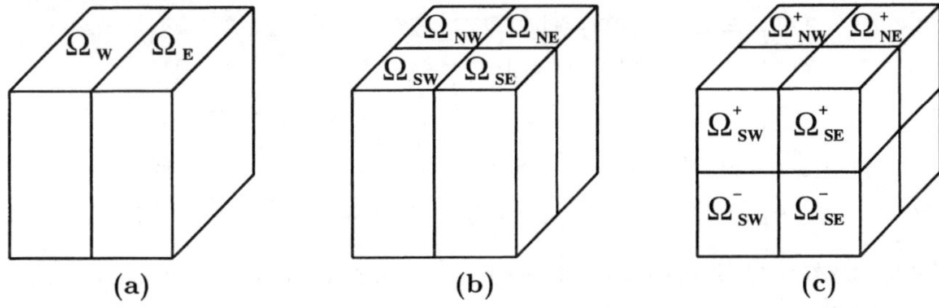

Fig. 4.14. Splitting of the cell Ω into eight subcells in three stages.

In the first stage, Ω is split into the two subcells

$$\Omega_W = [0..2^{p-1} - 1] \times [0..2^p - 1] \times [0..2^s - 1],$$
$$\Omega_E = [2^{p-1}..2^p - 1] \times [0..2^p - 1] \times [0..2^s - 1]$$

of equal size across the *i-axis* (Figure 4.14 **(a)**). The number $m_W = |\Omega_W|$ of pixels contained in the cell Ω_W is coded. These are at most $m = |\Omega|$ pixels. Therefore, for coding the number m_W we need $\lceil \log_2(m + 1) \rceil$ bits. Since the number $m_E = |\Omega_E|$ of pixels in Ω_E is given by $m_E = m - m_W$, we do not code m_E.

In the second stage, each of the two subcells Ω_W and Ω_E is split across the opposite *j-axis* (see Figure 4.14 **(b)**, yielding the four subcells

$$\begin{aligned}
\Omega_{SW} &= [0..2^{p-1}-1] \times [0..2^{p-1}-1] \times [0..2^s-1],\\
\Omega_{NW} &= [0..2^{p-1}-1] \times [2^{p-1}..2^p-1] \times [0..2^s-1],\\
\Omega_{SE} &= [2^{p-1}..2^p-1] \times [0..2^{p-1}-1] \times [0..2^s-1],\\
\Omega_{NE} &= [2^{p-1}..2^p-1] \times [2^{p-1}..2^p-1] \times [0..2^s-1].
\end{aligned}$$

The two numbers $m_{SW} = |\Omega_{SW}|$ and $m_{SE} = |\Omega_{SE}|$ are coded one after the other. Since the cell Ω_{SW} contains at most $m_W = |\Omega_W|$ pixels, we need $\lceil \log_2(m_W + 1) \rceil$ bits for coding m_{SW}. Likewise, coding the number m_{SE} requires $\lceil \log_2(m_E+1) \rceil$ bits. The two numbers m_{NW} and m_{NE} do not need to be coded. Indeed, this is because $m_{NW} = m_W - m_{SW}$ and $m_{NE} = m_E - m_{SE}$, i.e., the numbers m_{NW} and m_{NE} follow from the previous information in the bitstream.

In the third stage, each of the four subcells $\Omega_{SW}, \Omega_{NW}, \Omega_{SE}$, and Ω_{NE} is split across the k-*axis* (see Figure 4.14 **(c)**), each into two halves of equal size, which yields the eight subcells

$$\begin{aligned}
\Omega_{SW}^- &= [0..2^{p-1}-1] \times [0..2^{p-1}-1] \times [0..2^{s-1}-1],\\
\Omega_{NW}^- &= [0..2^{p-1}-1] \times [2^{p-1}..2^p-1] \times [0..2^{s-1}-1],\\
\Omega_{SE}^- &= [2^{p-1}..2^p-1] \times [0..2^{p-1}-1] \times [0..2^{s-1}-1],\\
\Omega_{NE}^- &= [2^{p-1}..2^p-1] \times [2^{p-1}..2^p-1] \times [0..2^{s-1}-1],\\
\Omega_{SW}^+ &= [0..2^{p-1}-1] \times [0..2^{p-1}-1] \times [2^{s-1}..2^s-1],\\
\Omega_{NW}^+ &= [0..2^{p-1}-1] \times [2^{p-1}..2^p-1] \times [2^{s-1}..2^s-1],\\
\Omega_{SE}^+ &= [2^{p-1}..2^p-1] \times [0..2^{p-1}-1] \times [2^{s-1}..2^s-1],\\
\Omega_{NE}^+ &= [2^{p-1}..2^p-1] \times [2^{p-1}..2^p-1] \times [2^{s-1}..2^s-1],
\end{aligned}$$

whose union is Ω. The four numbers $m_{SW}^- = |\Omega_{SW}^-|$, $m_{NW}^- = |\Omega_{NW}^-|$, $m_{SE}^- = |\Omega_{SE}^-|$, and $m_{NE}^- = |\Omega_{NE}^-|$ are coded.

Altogether, by splitting the bounding cell Ω into eight subcells, the sequence

$$m_W \,|\, m_{SW} \,|\, m_{SE} \,|\, m_{SW}^- \,|\, m_{NW}^- \,|\, m_{SE}^- \,|\, m_{NE}^-$$

of seven numbers is coded. This requires

$$\begin{aligned}
&\lceil \log_2(m+1) \rceil + \lceil \log_2(m_W+1) \rceil + \lceil \log_2(m_E+1) \rceil\\
&+\lceil \log_2(m_{SW}+1) \rceil + \lceil \log_2(m_{NW}+1) \rceil\\
&+\lceil \log_2(m_{SE}+1) \rceil + \lceil \log_2(m_{NE}+1) \rceil
\end{aligned}$$

bits in total, to be appended to the bitstream. The hierarchical structure of this coding scheme is shown in Figure 4.15.

This splitting (including the updates in the bitstream) is then recursively applied to those subcells which are not empty. A cell $\omega \subset \Omega$ is said to be *empty*, iff it contains no pixel, and thus $|\omega| = 0$. On the most elementary recursion level, we encounter cells of the form $\omega = [2i, 2i+1] \times [2j, 2j+1] \times [k]$,

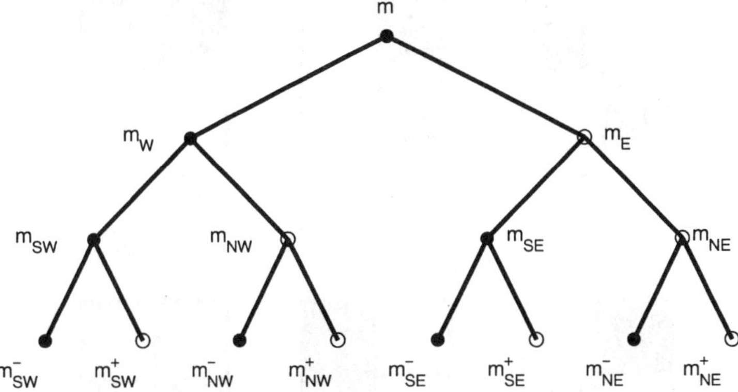

Fig. 4.15. Splitting of the domain Ω into eight subdomains $\Omega_{SW}^-, \Omega_{NW}^-, \Omega_{SE}^-, \Omega_{NE}^-, \Omega_{SW}^+, \Omega_{NW}^+, \Omega_{SE}^+, \Omega_{NE}^+$. The numbers $m, m_W, m_{SW}, m_{SE}, m_{SW}^-, m_{NW}^-, m_{SE}^-, m_{NE}^-$ (whose nodes are marked by •) are coded.

provided that $s < p$. Such *atomic* cells are not split. This is in order to save additional costs in terms of bits. Instead of this, the coding of the pixels in atomic cells is accomplished as follows.

We merely discuss the coding of the atomic cell $\omega = [0,1] \times [0,1] \times [0]$, all the other atomic cases are treated in an analogous manner. Note that the atom $\omega = [0,1] \times [0,1] \times [0]$ may contain zero, one, two, three or four pixels. In case ω contains either four or zero pixels, no additional information (in terms of coding costs) is required. If ω contains exactly one pixel, then its position $(i,j) \in \{0,1\} \times \{0,1\}$ is coded by using two bits, one for the index i and one for the index j. Likewise, if ω contains three pixels, then the other position in ω (which contains no pixel) is coded by using two bits. In the remaining case, where ω contains exactly two pixels, there are six different possibilities for the distribution of the two pixels in ω. These six different cases are coded according to the *Huffman code* $(1,1), (1,0), (0,0,1), (0,0,0), (0,1,1), (0,1,0)$, see Figure 4.16.

This requires only $2/6 \times 2$ bits $+ 4/6 \times 3$ bits $= 8/3$ bits in average, provided that the probabilities for each of the six cases are equal. This is cheaper than coding each of the two pixels separately, which would require 2×2 bits $= 4$ bits.

4.6.5 Computational Complexity

In this subsection, we analyze the computational complexity of the proposed image compression scheme. To this end, we determine the computational costs required for the performance of the steps **(1)**,**(2)**,**(4)**, and **(5)** at the outset in Subsection 4.6.1. Recall that step **(1)** is done by using adaptive thinning. But the complexity of the adaptive thinning is already discussed

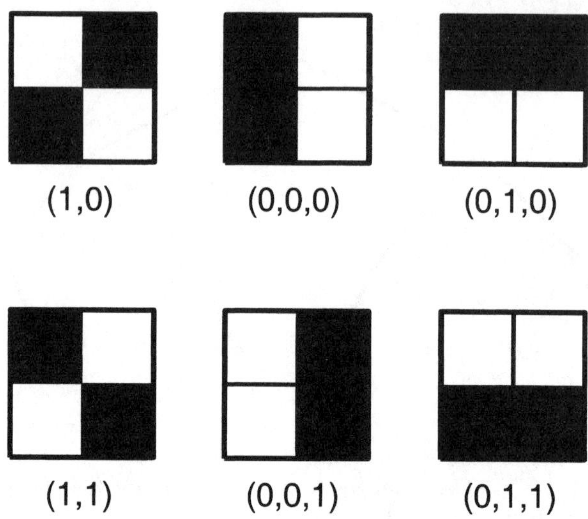

Fig. 4.16. Coding of the atomic domain $\Omega = [0..1] \times [0..1] \times [0]$ with two pixels, corresponding to the six different codes $(1,1), (1,0), (0,0,1), (0,0,0), (0,1,1), (0,1,0)$.

in Subsection 4.5.3. Recall from Subsection 4.5.3 that we require only at most $\mathcal{O}(N \log(N))$ operations for the removal of n pixels from a number of $N = 2^p \times 2^p$ pixels, when using any of the thinning algorithms **AT1**, **AT2**, or **AT3**. It is straightforward to establish the same complexity $\mathcal{O}(N \log(N))$ for the adaptive thinning algorithms $\mathbf{AT_2}$ and $\mathbf{AT_2^2}$, by following along the lines of the analysis in Subsection 4.5.3. In order to avoid unnecessary detours, however, we prefer to refrain from doing so here.

Instead, let us proceed by turning to the complexity of step **(5)**. In this step, the Delaunay triangulation $\mathcal{D}_Y$ is first constructed from the $m = N - n$ most significant pixel positions, before the corresponding piecewise linear function $\tilde{L}(Y; \tilde{F}_Y)$ is used in order to compute the n luminance values at the deleted pixel positions. Recall that building the Delaunay triangulation $\mathcal{D}_Y$ costs $\mathcal{O}(m \log(m))$ operations [134]. The subsequent reconstruction of the n luminance values costs $\mathcal{O}(n)$ operations, which is $\mathcal{O}(N)$ for large n.

As to the remaining two steps, **(2)** and **(4)**, note that these are symmetric. In fact, the asymptotic complexity of the encoding in step **(2)** is the same as the asymptotic complexity of the decoding in step **(4)**.

Therefore, we restrict ourselves to the analysis of the computational costs required for the performance of the encoding. To this end, recall from Subsection 4.6.4 that step **(2)** relies on the recursive splitting of the bounding cell Ω, containing $m = |\Omega|$ pixels. Therefore, we need to determine the computational costs required for the construction of the entire octtree-like data structure. Now note that the initial splitting of Ω into the two subcells Ω_W and Ω_E costs m operations in the first stage. Indeed, these m operations are required for counting the number m_W of pixels in Ω_W. But the subsequent splitting of Ω_W and Ω_E in the second stage costs also m operations, namely m_W for splitting Ω_W and m_E for splitting Ω_E, and so altogether $m_W + m_E = m$. By recursion, the splitting at each level ℓ costs exactly m operations. Now since the tree comprises $2p + s = \log_2(N) + s$ levels, this leads to $m \times (\log_2(N) + s) = \mathcal{O}(m \log(N))$ operations for the performance of step **(2)**.

Altogether, this shows that we require asymptotically at most

$$\mathcal{O}(N \log(N)) + \mathcal{O}(m \log(N)) = \mathcal{O}(N \log(N))$$

operations for the compression, in steps **(1)** and **(2)**, and only at most

$$\mathcal{O}(m \log(N)) + \mathcal{O}(m \log(m)) + \mathcal{O}(N) = \mathcal{O}(m \log(N)) + \mathcal{O}(N)$$

operations for the decompression in steps **(4)** and **(5)**.

4.6.6 Adaptive Thinning versus SPIHT

Now let us turn to the evaluation of adaptive thinning in image compression. In this subsection, we apply the adaptive thinning algorithms $\mathbf{AT_2}$ and $\mathbf{AT_2^2}$. We work with greyscale values of the luminances $f(i,j)$ in $[0..255]$, i.e., $r = 8$. In the test examples below, we use the quantization step 8, so that $[0..31]$ is the range of the quantized symbols $Q(f^*(i,j))$, $(i,j) \in Y$.

We compare the performance of our compression scheme with that of the wavelet-based compression scheme *Set Partitioning Into Hierarchical Trees* (SPIHT) [145]. We remark that the good compression rate of SPIHT is, at low bit rates, comparable with that of the powerful method *EBCOT* [166], which is the basis algorithm of the standard *JPEG2000* [167].

In order to compare our compression scheme with SPIHT, we use several different test images. In each test case, the compression rate, measured in *bits per pixel* (bpp), is fixed. The quality of the resulting reconstruction is then evaluated by the comparison of the differences in PSNR.

Geometric Test Images. We first consider two artificial test images, `Chessboard` and `Reflex`, each of small size 128×128 ($p = q = 7$). These two geometric test images are displayed in Figure 4.17 **(a)** and Figure 4.18 **(a)**.

The purpose of this preliminary discussion is two-fold. Firstly, we wish to demonstrate the good performance of adaptive thinning for texture-free images with sharp edges. Secondly, we demonstrate that the greedy two-point removal strategy of $\mathbf{AT_2^2}$ is superior to the greedy one-point removal strategy of $\mathbf{AT_2}$, and we provide a few arguments to explain it.

A first comparison between SPIHT, $\mathbf{AT_2}$ and $\mathbf{AT_2^2}$ is done by using the test image `Chessboard` in Figure 4.17 **(a)**. In this example, we considered selecting the 299 most significant pixels by using $\mathbf{AT_2^2}$ and $\mathbf{AT_2}$, respectively.

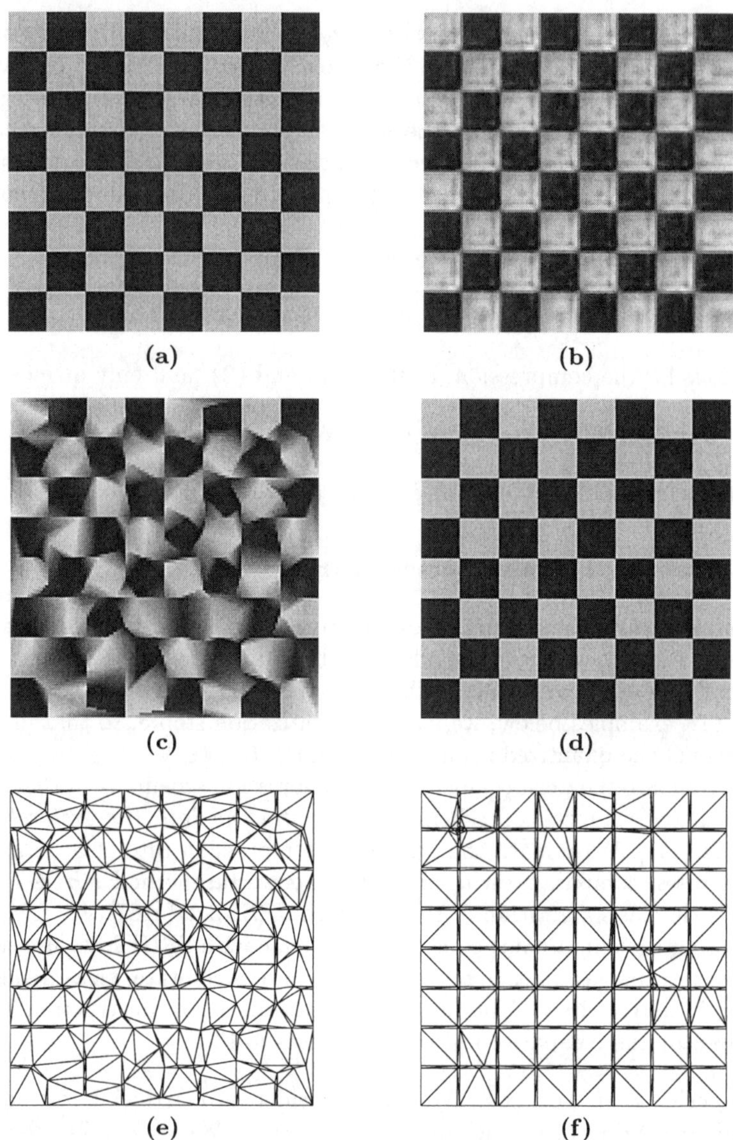

Fig. 4.17. Chessboard. (a) Original image of size 128 × 128. Reconstruction at 0.23 bpp by (b) SPIHT with PSNR 18.57 db, (c) $\mathbf{AT_2}$ with PSNR 15.24 db, (d) $\mathbf{AT_2^2}$ with PSNR 45.15 db. The Delaunay triangulations of the 299 most significant pixels output by adaptive thinning are shown in (e) for $\mathbf{AT_2}$ and in (f) for $\mathbf{AT_2^2}$.

In Figure 4.17, the Delaunay triangulations of the 299 most significant pixel positions, together with the resulting reconstructed images, are shown.

The algorithm $\mathbf{AT}_2^2$ selects an *optimal* subset Y of 299 most significant pixels, so that the least squares error in (4.42) is zero, namely $L(F, \mathcal{D}_Y)(i,j) = f(i,j)$ for all $(i,j) \in X$. Only by the quantization of the luminance values, the resulting mean square error for the reconstructed image $\tilde{F}$ is slightly increased to $\bar{\eta}_2^2(Y;X) = 2.002$, which corresponds to a PSNR of 45.15 dB. The almost exact reconstruction of the original image Chessboard, provided by $\mathbf{AT}_2^2$, is shown in Figure 4.17 (**d**).

In contrast, SPIHT leads to an inferior PSNR of only 18.57 dB, and the algorithm $\mathbf{AT}_2$ leads to a PSNR of 15.24 dB. Moreover, the quality of their resulting reconstruction is rather poor, see Figure 4.17 (**b**) for the reconstruction of SPIHT and Figure 4.17 (**c**) for $\mathbf{AT}_2$. We have recorded the results of this example, along with those of the following test cases, in Table 4.2.

We can explain the superiority of $\mathbf{AT}_2^2$ over $\mathbf{AT}_2$ for the test case Chessboard as follows. First of all, the algorithm $\mathbf{AT}_2^2$ looks two removal steps ahead, whereas $\mathbf{AT}_2$ looks only one step ahead. Therefore, the algorithm $\mathbf{AT}_2^2$ allows the removal of edges $[y_1, y_2] \in \mathcal{D}_Y$, whose corresponding anticipated error $e(y_1, y_2)$ is small, even if the anticipated errors $e(y_1)$ and $e(y_2)$ may be large. This is typically the case for pixels y_1, y_2 with an edge $[y_1, y_2]$, crossing the boundary between two squares of the chessboard in a nearly perpendicular direction, away from the corners.

In this case, the algorithm $\mathbf{AT}_2^2$ removes the *non-significant* edge $[y_1, y_2] \in \mathcal{D}_Y$, whereas the algorithm $\mathbf{AT}_2$ is too short-sighted to make such removals, and so $\mathbf{AT}_2$ prefers to keep both y_1 and y_2. This leads to an early removal of pixels near the corners of the chessboard squares, see Figure 4.17 (**e**). In contrast, the algorithm $\mathbf{AT}_2^2$ manages to keep pixels near the corners, see Figure 4.17 (**f**).

Now let us turn to the other geometric test image, Reflex, displayed in Figure 4.18 (**a**). In this test case, we fix the compression rate to 0.251 bpp. The resulting reconstruction from $\mathbf{AT}_2^2$ is, in comparison with that of SPIHT, displayed in Figure 4.18 (**b**),(**d**). Our method $\mathbf{AT}_2^2$ yields the PSNR value 42.86 dB, whereas SPIHT provides the inferior PSNR value 30.42 dB. Hence, with respect to this quality measure, our method is much better. Moreover, the reconstruction by $\mathbf{AT}_2^2$ provides also a superior *visual quality* to that of the reconstructed image by SPIHT, see Figures 4.18 (**b**),(**d**). Indeed, $\mathbf{AT}_2^2$ manages to localize the sharp edges of the test image Reflex. Moreover, it avoids undesired oscillations around the edges, unlike SPIHT. This is due to the well-adapted distribution of the 384 most significant pixels, whose Delaunay triangulation is displayed in Figure 4.18 (**c**).

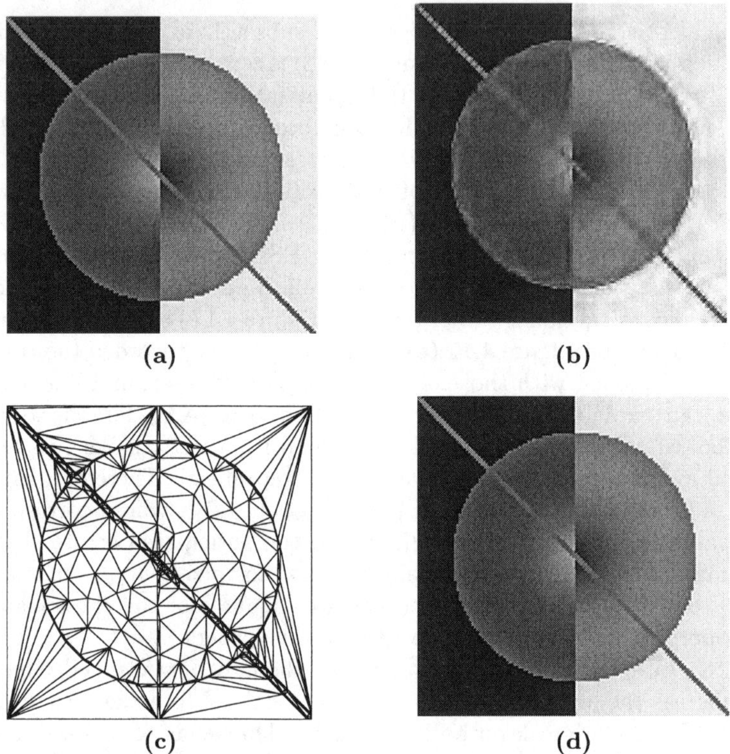

(a) (b) (c) (d)

Fig. 4.18. `Reflex`. **(a)** Original image of size 128×128. Reconstruction at 0.251 `bpp` by **(b)** SPIHT with PSNR 30.42 `db`, **(d)** $\mathbf{AT}_2^2$ with PSNR 42.86 `db`. The Delaunay triangulations of the 384 most significant pixels output by $\mathbf{AT}_2^2$ is shown in **(c)**.

Three Popular Test Cases of Real Images. We considered also applying the adaptive thinning algorithms $\mathbf{AT}_2$ and $\mathbf{AT}_2^2$ on three different popular test cases from image processing, `Fruits`,`Peppers` and `Lena`, which are also used as standard test cases in the textbook [167]. The original images, each of size 512 × 512, are displayed in Figures 4.21, 4.24, and 4.27.

Not surprisingly, we found that the algorithm $\mathbf{AT}_2^2$ is superior to the algorithm $\mathbf{AT}_2$. It is remarkable that the algorithm $\mathbf{AT}_2^2$ is, at low bitrates, quite competitive with SPIHT. This is confirmed by the results in Figure 4.19, where the PSNR as a function of the compression rate (in bits per pixel) is plotted for the three different compression algorithms, SPIHT, $\mathbf{AT}_2$ and $\mathbf{AT}_2^2$.

In the following discussion on the *visual quality* of the reconstruction, we focus on the comparison between SPIHT and $\mathbf{AT}_2^2$. For each of the three test images, the different PSNR values are shown in Table 4.2. Note that the PSNR obtained by $\mathbf{AT}_2^2$ is slightly larger than that obtained by SPIHT for the two test cases `Fruits` and `Peppers`, but slightly inferior for the test case `Lena`.

Now let us turn to the visual quality of the reconstructions. The reconstruction by SPIHT and $\mathbf{AT}_2^2$, respectively, is shown in the Figures 4.23, 4.26, and 4.29.

The set Y of most significant pixel positions, along with its Delaunay triangulation $\mathcal{D}_Y$, are displayed in Figure 4.20. Note that by the distribution of the most significant points, the main features of the images, such as sharp edges and silhouettes, are captured very well, see Figures 4.22,4.25, and 4.28. Moreover, the compression scheme $\mathbf{AT}_2^2$ manages to denoise the test image `Fruits` quite successfully, in contrast to SPIHT.

On balance, in terms of the *visual* quality of the reconstructions of `Fruits`, `Peppers`, and `Lena`, we believe that our compression method $\mathbf{AT}_2^2$ is just as good as SPIHT. This is widely supported by the results in Table 4.2 and Figure 4.19.

Table 4.2. Comparison between SPIHT, $\mathbf{AT}_2^2$, and $\mathbf{AT}_2$

			PSNR	
Test Case	**bpp**	SPIHT	$\mathbf{AT}_2^2$	$\mathbf{AT}_2$
Chessboard	0.230	18.57	45.15	15.24
Reflex	0.251	30.42	42.86	41.73
Fruits	0.185	32.33	32.37	31.85
Peppers	0.158	32.07	32.30	31.80
Lena	0.154	32.26	31.50	30.98

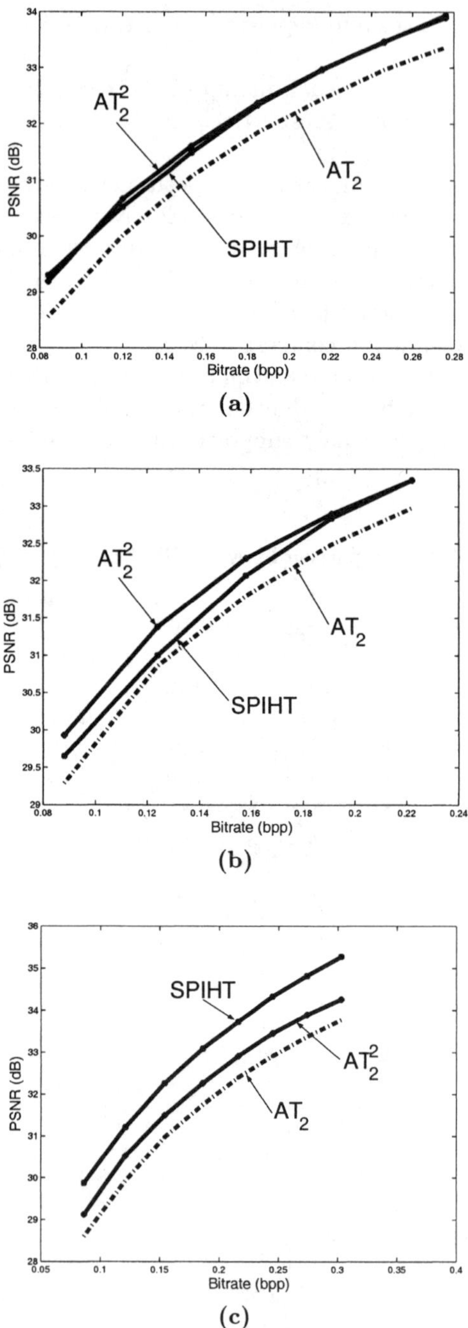

Fig. 4.19. Comparison between SPIHT, $\mathbf{AT_2}$ and $\mathbf{AT_2^2}$ at low bitrates for the test images (a) `Fruits`, (b) `Peppers`, and (c) `Lena`.

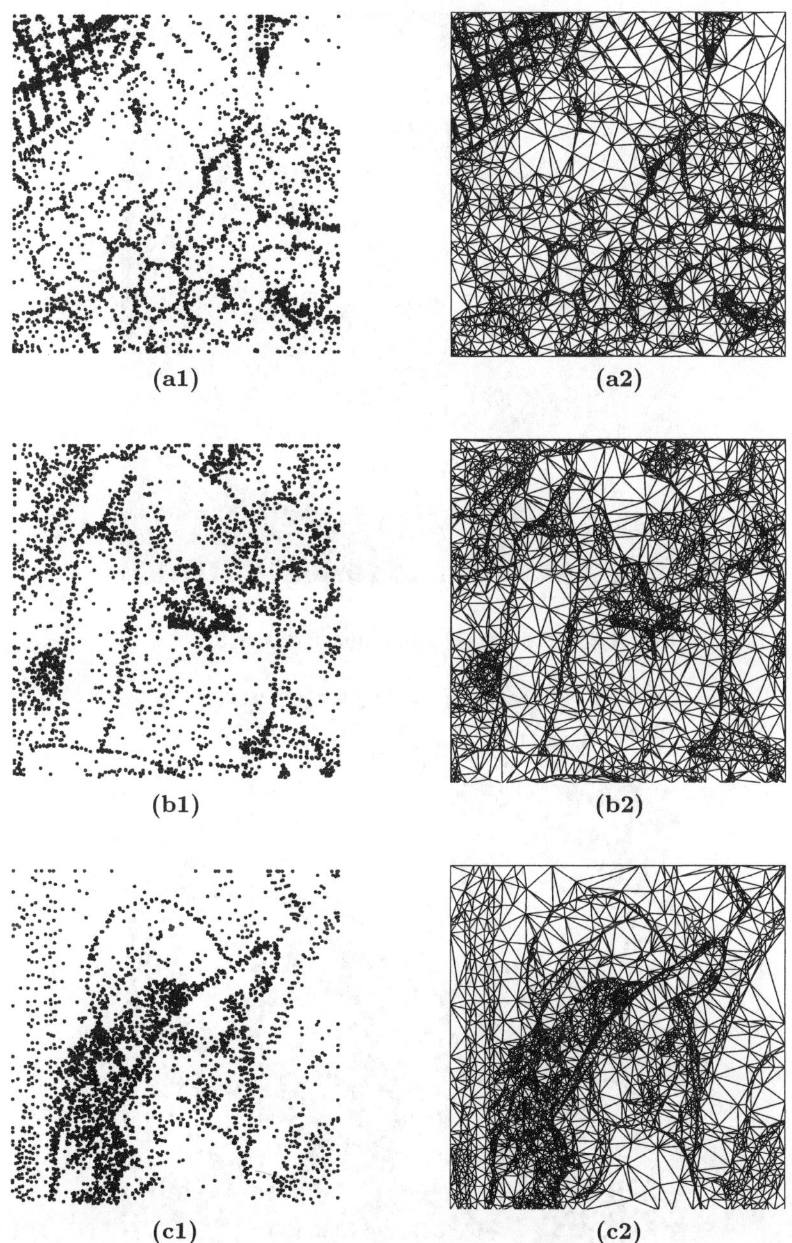

Fig. 4.20. Most significant pixel positions and their Delaunay triangulation, selected by $\mathbf{AT}_2^2$ for the test images **(a1)**,**(a2)** `Fruits`, **(b1)**,**(b2)** `Peppers`, and **(c1)**,**(c2)** `Lena`.

Fig. 4.21. `Fruits`. Original image of size 512 × 512.

Fig. 4.22. `Fruits`. 4044 most significant pixels selected by $\mathbf{AT}_2^2$.

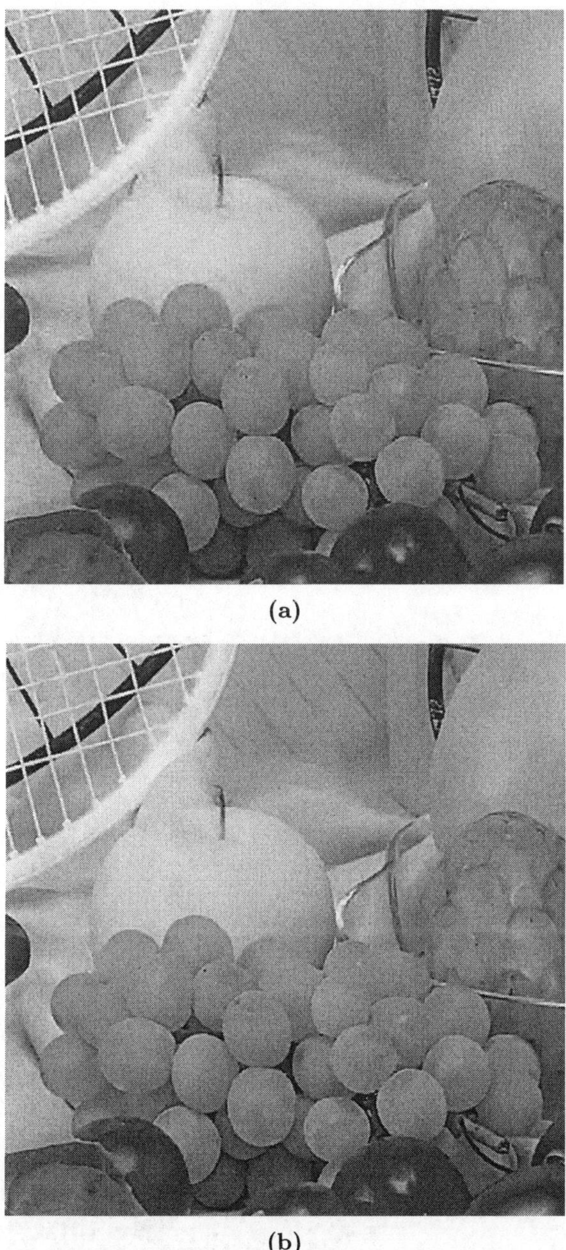

(a)

(b)

Fig. 4.23. `Fruits`. Reconstruction at 0.185 bpp by **(a)** SPIHT with PSNR 32.33 db and **(b)** $\mathbf{AT}_2^2$ with PSNR 32.37 db.

Fig. 4.24. `Peppers`. Original image of size 512 × 512.

Fig. 4.25. `Peppers`. 3244 most significant pixels selected by $\mathbf{AT}_2^2$.

(a)

(b)

Fig. 4.26. `Peppers`. Reconstruction at 0.158 bpp by **(a)** SPIHT with PSNR 32.07 db and **(b)** $\mathbf{AT}_2^2$ with PSNR 32.30 db.

Fig. 4.27. Lena. Original image of size 512 × 512.

Fig. 4.28. Lena. 3244 most significant pixels selected by $\mathbf{AT}_2^2$.

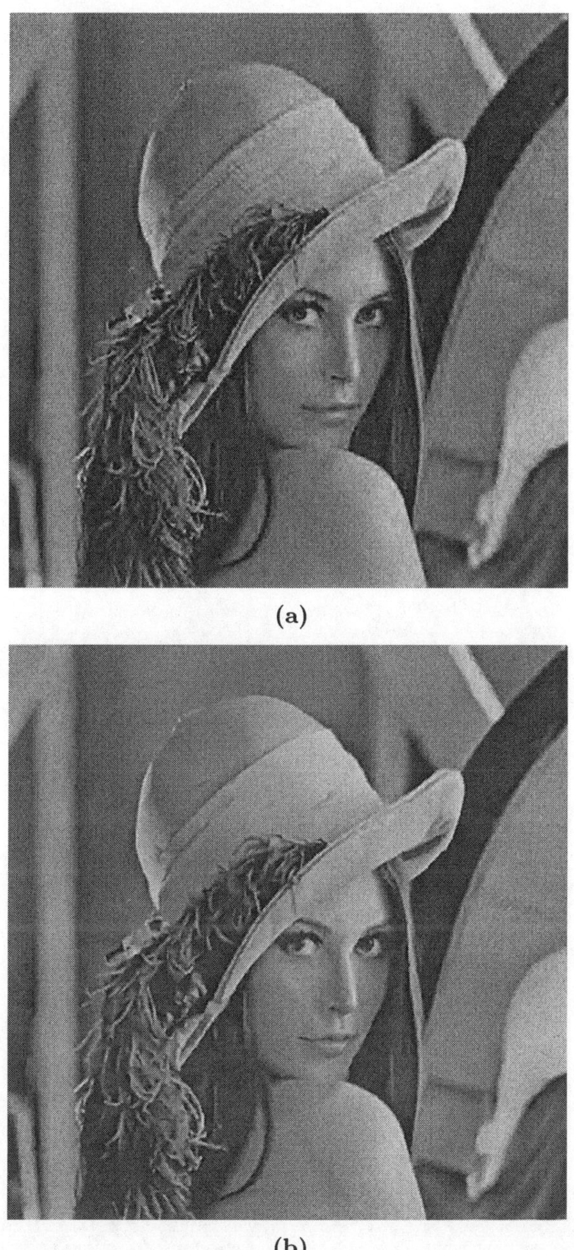

(a)

(b)

Fig. 4.29. Lena. Reconstruction at 0.154 bpp by (a) SPIHT with PSNR 32.26 db and (b) $\mathbf{AT}_2^2$ with PSNR 31.50 db.

5 Multilevel Approximation Schemes

Multilevel approximation schemes are concerned with the construction of a hierarchical representation of a model object, a mathematical function, at various different resolutions. This chapter first reviews our recent and current research on multilevel approximation from scattered data, before special emphasis is placed on their application to hierarchical surface visualization.

Starting point of the discussion in this chapter is our multilevel interpolation scheme of [72], which was the first to combine thinning algorithms with scattered data interpolation by compactly supported radial basis functions. As confirmed by the extensive numerical examples in [73], the multilevel interpolation scheme of [72] is very efficient and stable. Moreover, its utility in real-world model problems and industrial applications, such as in geometric modelling and subsurface recovery from seismic data, is demonstrated in [73]. We remark that the multilevel concept of [72] is utilized, further discussed, and (partly) analyzed in related papers [63, 64, 65, 66, 126, 171].

Various improvements of the multilevel interpolation scheme of [72] are introduced in [96], where it is shown that the performance of the multilevel interpolation heavily relies on the required preprocessing data analysis, which is in [72] done by using non-adaptive thinning, see Section 4.3. The improvements in [96] are basically achieved by replacing non-adaptive thinning with scattered data filtering, see Section 4.4, in the performance of the preprocess.

Just very recently, the multilevel concept of [72, 96] was further improved in our paper [98], where we combine adaptive domain decomposition and data clustering with stable local polyharmonic spline interpolation, see Subsection 3.8.3. This yields enhanced flexibility, when it comes to combining the required data analysis with the data synthesis.

In the following Section 5.1, a generic formulation for multilevel scattered data approximation is provided, which unifies the various multilevel schemes mentioned above. Section 5.2 is then devoted to multilevel interpolation, as suggested in [72], and further developed in [96], before the recent multilevel approximation scheme of [98] is explained in Section 5.3. In Section 5.4, the behaviour of the different multilevel approximation schemes is evaluated and compared by using one model problem from hierarchical surface visualization.

5.1 Generic Formulation

Starting point for any multilevel approximation scheme, discussed in this chapter, is the construction of a data hierarchy of the form

$$X_1 \subset X_2 \subset \cdots \subset X_{L-1} \subset X_L = X \tag{5.1}$$

from given sample values $f\big|_X = (f(x_1), \ldots, f(x_N))^T \in \mathbb{R}^N$ taken from an unknown function $f : \mathbb{R}^d \to \mathbb{R}$ at a set $X = \{x_1, \ldots, x_N\} \subset \mathbb{R}^d$ of scattered locations. The above data hierarchy (5.1) may, for instance, be constructed a priori by using one of the thinning algorithms, as discussed in the previous chapter. An alternative construction technique is working with adaptive domain decomposition, as discussed later in this chapter. In either case, this construction process is referred to as the *data analysis.*

Now the basic idea of multilevel approximation is to first approximate f from the samples $f\big|_{X_1}$ taken at the *coarsest* level by computing an approximation s_1 which captures the global trend of the function f. In the subsequent steps of the multilevel scheme, a sequence $s_1, \ldots, s_L$ of gradually finer representations for f is constructed by computing a sequence $\Delta s_1, \ldots, \Delta s_L$ of approximations to the residual of the previous level. More precise, at each level ℓ, $1 \le \ell \le L$, the approximation Δs_ℓ to the residual $(f - s_{\ell-1})\big|_{X_\ell}$ yields by $s_\ell = s_{\ell-1} + \Delta s_\ell$ a *finer* representation for f (compared with the *coarser* one, $s_{\ell-1}$, of the previous level).

Altogether, the construction of the multiresolution representation for f, called the *data synthesis*, is accomplished by using the following algorithm.

Algorithm 18 (Multilevel Approximation).

(1) *Let* $s_0 \equiv 0$*;*
(2) **FOR** $\ell = 1, \ldots, L$
  (2a) *Compute approximation* Δs_ℓ *to the data* $(f - s_{\ell-1})\big|_{X_\ell}$*;*
  (2b) *Let* $s_\ell = s_{\ell-1} + \Delta s_\ell$*;*

OUTPUT: *Sequence of approximations* $s_1, \ldots, s_L$ *to* f.

Especially in situations where the number $|X| = N$ of sample points is extremely large and the sampling density in the point set X is subject to strong variations, multilevel scattered data approximation schemes, are appropriate tools. This is mainly because, due to its heterogeneity, such data sets naturally incorporate multiple resolutions. Therefore, when it comes to the data modelling, it makes sense to select a *multiresolution method*, such as Algorithm 18, in order to represent the function f at different resolutions.

In order to determine a specific multilevel approximation scheme from Algorithm 18, it remains to specify a particular approximation scheme, yielding a sequence $\mathcal{S}_1, \ldots, \mathcal{S}_L$ of approximation spaces, so that at any level ℓ the function Δs_ℓ, computed in step **(2a)** of Algorithm 18, is an element of $\mathcal{S}_\ell$.

In order to make concrete examples, the approximation scheme in could, for instance, employ piecewise linear interpolation over triangulations. In this case, it makes sense to select, at level ℓ, the space $\mathcal{S}_\ell = S_1^0(\mathcal{T}_\ell)$, being the linear finite elements over a fixed triangulation $\mathcal{T}_\ell$ of X_ℓ.

One could also work with least squares approximation or plain interpolation by using radial basis functions, as explained in the Sections 3.1 and 3.10. In either case, it makes sense to select finite-dimensional linear approximation spaces of the form

$$\mathcal{S}_\ell \equiv \mathcal{S}_{\phi_\ell, X_\ell} = \text{span}\left\{\phi_\ell(\|\cdot - x\|) \,:\, x \in X_\ell\right\} \bigoplus \mathcal{P}_{m_\ell}^d, \quad 1 \leq \ell \leq L, \tag{5.2}$$

each of whose major part is spanned by the X_ℓ-translates of a specific radial basis function $\phi_\ell \in \mathbf{CPD}_d(m_\ell)$, see the form of the recovery space $\mathcal{R}_{\phi,Y}$ in Section 3.10.

5.2 Multilevel Interpolation

In this section, we discuss multilevel interpolation, as first proposed in the paper [72]. The scheme utilizes radial basis function interpolation, and so it works with approximation spaces $\mathcal{S}_1, \ldots, \mathcal{S}_L$ of the form (5.2). We remark that, despite that the data hierarchy of the points in (5.1) is nested, these approximation spaces are not necessarily nested. This is in contrast to wavelets, whose theory of *multiresolution analysis* (MRA), essentially relies on the nestedness of consecutive *wavelet spaces.* The construction of such wavelet spaces is based on the *refinement* of the *wavelet* (basis) function. Due to the lack of such refinement equations, radial basis functions do not work with nested approximation spaces.

Now let us first formulate the multilevel interpolation algorithm, before its major ingredients are discussed below.

Algorithm 19 (Multilevel Interpolation).

INPUT: *Data hierarchy* $X_1 \subset \cdots \subset X_L = X$, *and function values* $f\big|_X$.

(1) *Let* $s_0 \equiv 0$;
(2) FOR $\ell = 1, \ldots, L$
  (2a) *Compute interpolant* $\Delta s_\ell \in \mathcal{S}_\ell$ *satisfying* $(f - s_{\ell-1})\big|_{X_\ell} = \Delta s_\ell\big|_{X_\ell}$;
  (2b) *Let* $s_\ell = s_{\ell-1} + \Delta s_\ell$;

OUTPUT: *Sequence of interpolants* $s_1, \ldots, s_L$ *to* f.

The basic idea of this multilevel interpolation scheme is to first capture a global trend of the function f (corresponding to low frequencies of f) at the initial level $\ell = 1$, before finer details (corresponding to higher frequencies of f) are gradually added by *local* updates in the subsequent step **(2b)**. The resulting interpolants $s_1, \ldots, s_L$ provide a sequence of representations for f

at L different resolutions. To this end, in the above multilevel interpolation scheme, the following L interpolation problems are to be solved one after the other.

$$\begin{aligned} f\big|_{X_1} &= \Delta s_1\big|_{X_1}, \\ (f - s_1)\big|_{X_2} &= \Delta s_2\big|_{X_2}, \\ &\vdots \quad \vdots \\ (f - s_{L-1})\big|_{X_L} &= \Delta s_L\big|_{X_L}. \end{aligned} \tag{5.3}$$

Note that every function s_ℓ in (5.3) matches f at the subset X_ℓ, i.e.,

$$f\big|_{X_\ell} = s_\ell\big|_{X_\ell}, \qquad \text{for all } 1 \leq \ell \leq L. \tag{5.4}$$

As to the selection of the radial basis functions ϕ_ℓ of the approximation spaces $\mathcal{S}_\ell$, $1 \leq \ell \leq L$, our recommendation is to work with polyharmonic splines at level one, where $\ell = 1$. This is due to the useful variational principle for polyharmonic splines (as discussed more detailed in Section 3.8), ensuring a good global approximation behaviour of the resulting surface spline. Hence, the initial interpolant s_1 to f in (5.3) has the form

$$s_1 = \sum_{x \in X_1} c_x \phi_{d,k}(\| \cdot - x \|) + p_k, \tag{5.5}$$

with $p_k : \mathbb{R}^d \to \mathbb{R}$ being a polynomial of order at most k, cf. the representation for a polyharmonic spline interpolant in (3.44). Recall from the discussion in Section 3.8 of Chapter 3 that polyharmonic spline interpolation, using $\phi_{d,k}$, reproduces any element from the linear space $\mathcal{P}_k^d$, comprising all d-variate polynomials of order at most k, provided that the points in X_1 are $\mathcal{P}_k^d$-unisolvent. Hence, when working with the thin plate splines in the plane, where $d = k = 2$ and thus $\phi_{2,2} = r^2 \log(r)$, the interpolation scheme reproduces linear polynomials, provided that the points in X_1 do not all lie on a straight line.

It is easy to see that this reproduction property carries over to the multilevel interpolation scheme of Algorithm 19. Indeed, in this case, we obtain, for any polynomial $f \in \mathcal{P}_k^d$, the identity $s_1 = \Delta s_1 \equiv f$ when solving the initial interpolation problem at the coarsest level, cf. the first line in (5.3). But this implies that all subsequent interpolants Δs_ℓ, $2 \leq \ell \leq L$, in (5.3) vanish identically. Let us make a note of this simple but useful observation in a separate remark.

Remark 1. The multilevel interpolation scheme of Algorithm 19 reproduces any polynomial from $\mathcal{P}_k^d$, provided that the interpolation scheme utilized at its initial level reproduces polynomials from $\mathcal{P}_k^d$.

As to the interpolation at the remaining levels $2 \leq \ell \leq L$, we prefer to work with *compactly supported* radial basis functions, with using different scales at different levels. In this case, for a fixed compactly supported radial basis function ϕ with support radius $r = 1$ (see the selection in Table 3.2 on page 34), we let $\phi_\varrho = \phi(\cdot/\varrho)$ for $\varrho > 0$, and so any interpolant Δs_ℓ (used at level ℓ) in (5.3) has the form

$$\Delta s_\ell = \sum_{x \in X_\ell} c_x \phi_{\varrho_\ell}(\|\cdot - x\|), \quad 2 \leq \ell \leq L, \tag{5.6}$$

where the support radii ϱ_ℓ are a monotonically decreasing sequence of positive numbers, i.e., $\varrho_2 > \varrho_3 > \ldots > \varrho_L > 0$.

Note that the resulting multilevel interpolation scheme performs, at any level ℓ, $2 \leq \ell \leq L$, in step **(2b)** of Algorithm 19 merely a *local* update on the previous *coarser* interpolant $s_{\ell-1}$, yielding the *finer* interpolant s_ℓ to f. In order to ensure that these updates make sense, each support radius ϱ_ℓ should depend on the density of the set X_ℓ, $2 \leq \ell \leq L$, in the domain Ω. We can explain this as follows.

On the one hand, the choice of the support radius ϱ_ℓ allows us to steer the approximation quality of the update at level ℓ. For the sake of approximation quality, it is desirable to work a large support radius ϱ_ℓ. On the other hand, these updates should be local, and therefore we wish to keep the support radius ϱ_ℓ *reasonably* small. Moreover, a small ϱ_ℓ increases the *sparsity* of the arising (symmetric and positive definite) collocation matrix A_{ϕ_ℓ, X_ℓ}, and thus leads to a stable interpolation process. In fact, this is the dilemma of the uncertainty principle for radial basis functions, which is discussed more detailed in Section 3.7 of Chapter 3.

Now in order to be able to balance these conflicting requirements, the approximation quality and the stability (sparsity of A_{ϕ_ℓ, X_ℓ}) of the interpolation process, it is recommended to let the support radius ϱ_ℓ be proportional to the fill distance $h_{X_\ell,\Omega}$ of the point set X_ℓ in the domain Ω. This recommendation is based on our previous experience with multilevel interpolation by (compactly supported) radial basis functions in many different applications. E.g., in the situation of the small model problem in [72], we let $\varrho_\ell \approx 5 * h_{X_\ell,\Omega}$.

But the performance of this particular multilevel interpolation scheme does not solely depend on the choice of the support radii. In fact, the performance of the multilevel scheme relies heavily on the choice of the data hierarchy in (5.1). Therefore, considerable effort has gone into the design of algorithms for the construction of *suitable* data hierarchies of the form (5.1) [74, 96, 101]. These constructions rely mainly on thinning algorithms as well as on customized schemes for progressive scattered data filtering, as discussed in the previous Chapter 4.

The final Section 5.4 of this chapter is devoted to numerical comparisons concerning the performance of the multilevel interpolation scheme of this section, when working with different techniques for constructing the data hierarchy (5.1).

5.3 Adaptive Multilevel Approximation

In this section, a recent multilevel approximation scheme [98] is proposed. The scheme relies on an adaptive domain decomposition strategy using quadtree techniques (and their higher-dimensional generalizations), as explained in Section 2.6 of Chapter 2. It is shown in the numerical examples in Section 5.4 that this method achieves an improvement on the approximation quality of the well-established multilevel interpolation schemes of the previous Section 5.2 at the coarser levels.

In contrast to the multilevel scheme in Section 5.2, we drop the restriction of requiring the sets in (5.1) to be *subsets* of the given points X. Instead of this, although we work with a data hierarchy of the form

$$C_1 \subset C_2 \subset \cdots \subset C_{L-1} \subset C_L, \tag{5.7}$$

the sets C_ℓ in (5.7) do not necessarily need to be subsets of X. The locations of the points in the subsets C_ℓ, $\ell \geq 1$, are constructed, such that each point in C_ℓ represents a certain *cluster* in X. This essentially amounts to working with multilevel *approximation* rather than multilevel *interpolation*. In this new setting, each subset X_ℓ, used the interpolation step **(2a)** of Algorithm 19, is replaced by a corresponding set C_ℓ in (5.7). Moreover, in contrast to the situation in Algorithm 19, these sets C_ℓ are not given beforehand, but they are adaptively constructed during the performance of this more sophisticated multilevel scheme, given by the following Algorithm 20.

Algorithm 20 (Adaptive Multilevel Approximation).

INPUT: *Point set* X, *function values* $f\big|_X$.

(1) *Let* $s_0 \equiv 0$;
(2) *Construct the set* C_1;
(3) *Approximate* f *at* C_1, *and so obtain the data* $f\big|_{C_1}$;
(4) **FOR** $\ell = 1, 2, \ldots$
  (4a) *Compute interpolant* $\Delta s_\ell \in \mathcal{S}_\ell$ *satisfying* $(f - s_{\ell-1})\big|_{C_\ell} = \Delta s_\ell\big|_{C_\ell}$;
  (4b) *Let* $s_\ell = s_{\ell-1} + \Delta s_\ell$;
  (4c) *Construct the set* $C_{\ell+1}$ *satisfying* $C_\ell \subset C_{\ell+1}$;
  (4d) *Approximate* $f - s_\ell$ *at* $C_{\ell+1}$, *and so obtain the data* $(f - s_\ell)\big|_{C_{\ell+1}}$;

OUTPUT: *Sequence of approximations* $s_1, \ldots, s_L$ *to* f.

The construction of the data hierarchy (5.7) during the above Algorithm 20 is the subject of most of the following discussion in this section. Be it sufficient for the moment to say that the construction of the coarsest set C_1 in step **(2)** depends merely on the spatial distribution of the points in the given set X, but not on the sample values in $f\big|_X$. In contrast to this, the construction of any subsequent subset $C_{\ell+1}$ in step **(4c)** does essentially depend on the approximation behaviour of the current approximation s_ℓ, obtained in step **(4b)**. Details on the construction of the initial set C_1 and the subsequent sets $C_{\ell+1}$, $\ell \geq 1$, are explained below.

5.3.1 Adaptive Domain Decomposition

In this subsection, the construction of the sets in (5.7) is explained.

Let $\Omega \subset \mathbb{R}^d$ be a *bounding box* for X, i.e., Ω is a hypercuboid in $\mathbb{R}^d$ containing the point set X. In what follows, we intend to decompose Ω into a collection $\{\omega\}_{\omega \in \mathcal{L}}$ of smaller *cells* of different sizes and satisfying

$$\Omega = \bigcup_{\omega \in \mathcal{L}} \omega. \tag{5.8}$$

This yields by letting $X_\omega = X \cap \omega$, for $\omega \in \mathcal{L}$, a partition $\{X_\omega\}_{\omega \in \mathcal{L}}$ of X into accordingly many point clusters.

The decomposition of Ω (and thus the partition of X) will be computed by recursively splitting the cells, with Ω being the initial cell. This is done by using the data structure *quadtree* (and its higher-dimensional generalizations), as explained in Section 2.6. Each cell corresponds to a leaf in the (generalized) quadtree, and so initially, the cell Ω is contained in the root of the quadtree. The decision on the splitting of a single cell will be made according to two different customized splitting criteria, to be explained below. The cells are always split *uniformly* by using the function `split-cell` of Algorithm 14 in Subsection 2.6.1, respectively Algorithm 12 in Section 2.6 for the special case of two dimensions. The splitting of a 2D cell ω is shown in Figure 2.7 on page 27. Note that one could also consider *non-uniform* splittings of ω, but we want to avoid long and thin cells.

Splitting of Cells at the Coarsest Level. Initially, the cells are split according to the spatial distribution of the points in X. Our aim is to construct the *coarsest* set C_1 in (5.7). To this end, a decomposition (5.8) of Ω is computed, such that the number $|X_\omega|$ of points in each resulting point cluster X_ω is not greater than a predetermined number $n \ll |X| = N$.

Definition 22. *Let n be given. We say that a cell $\omega \in \mathcal{L}$ is* `splittable`*, iff the size $|X_\omega|$ of X_ω is greater than n, i.e., $|X_\omega| > n$.*

Having specified the definition for a splittable cell, the recursive domain decomposition of Ω is performed by applying the function `build-quadtree`, Algorithm 15 in Subsection 2.6.1 (respectively Algorithm 13 in Section 2.6 for the special case of two dimensions), on Ω.

Note that the call `build-quadtree(`Ω`)`, Algorithm 15, returns a partition $\{X_\omega\}_{\omega \in \mathcal{L}}$ of X, where each cluster X_ω contains not more than n points. Figure 5.1 shows an example, where this algorithm was applied on the data set Hurrungane (displayed in Figure 4.9, page 100). In this case, where $d = 2$, we have $|X| = 23092$ and $\Omega = [437000, 442000] \times [6812000, 6817000]$. For the decision concerning the splitting of a cell, the value $n = 60$ was selected. In this case, the function `build-quadtree` computes a decomposition of Ω into 841 cells, shown in Figure 5.1 **(a)**. Now this decomposition is used in order to define the coarsest set C_1 in (5.7). To this end, we let for any cell $\omega \in \mathcal{L}$ the point x_ω denote its center of gravity. We define the initial set C_1 by the union

$$C_1 = \{x_\omega \,:\, \omega \in \mathcal{L} \text{ and } X_\omega \text{ not empty }\} \tag{5.9}$$

of all cell centers whose cells are not empty, and so any $x_\omega \in C_1$ represents the point cluster $X_\omega \subset X$. In the situation of our example in Figure 5.1, one cell is empty. Therefore, the resulting set C_1, displayed in Figure 5.1 **(b)**, comprises $|C_1| = 840$ cell centers.

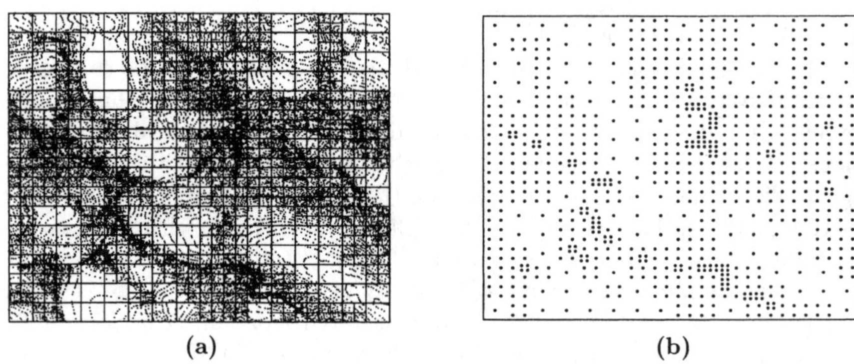

Fig. 5.1. Hurrungane. **(a)** Partitioning of the domain Ω into 841 cells $\{\omega\}_{\omega\in\mathcal{L}}$; **(b)** the centers x_ω of non-empty cells ω, yielding the subset C_1 of size $|C_1| = 840$.

Note that the points in C_1 do not have any function values, yet. To this end, in step **(3)** of Algorithm 20, we assign to each $x_\omega \in C_1$ a *cell average value* $s_\omega(x_\omega)$ of f on ω, where s_ω denotes a polyharmonic spline interpolant of the form (3.44) satisfying

$$s_\omega(x) = f(x), \qquad \text{for all } x \in X_\omega. \tag{5.10}$$

Having computed these cell average values, the interpolation in step **(4a)** of Algorithm 20 is, at the initial level, where $\ell = 1$, well-defined.

But we need to make a few comments concerning the numerical stability of this local polyharmonic spline interpolation. We first remark that the interpolation problem (5.10) is ill-posed, whenever the point set X_ω is not unisolvent with respect to the polynomials $\mathcal{P}_k^d$. In this situation, for the sake of numerical stability, we prefer to take the *mean value*

$$s_\omega(x_\omega) = \frac{1}{|X_\omega|} \sum_{x\in X_\omega} f(x)$$

for the cell average, rather than solving (5.10) by polyharmonic spline interpolation.

Moreover, we remark that the linear system resulting from (5.10) may be ill-conditioned, even if the interpolation problem (5.10) is itself well-conditioned. To be more precise, due to Narcowich and Ward [128], the spectral condition number of the resulting coefficient matrix is bounded above by a monotonically decreasing function of the interpolation points' *separation distance*

$$q_{X_\omega} = \min_{\substack{x,y \in X_\omega \\ x \neq y}} \|x - y\|,$$

see the discussion in Section 3.6. This in turn implies that one should, for the sake of numerical stability, avoid solving the linear system resulting from (5.10) directly in situations where the minimal distance q_{X_ω} between two points in X_ω is small. For further details on this, see [128] and the more general discussion in [151].

However, recall that in Subsection 3.8.3 a numerically stable algorithm for the evaluation of the interpolant s_ω in (5.10) is offered, see also [102]. The algorithm works with a rescaling of the interpolation points X_ω, so that their separation distance q_{X_ω} increases, and thus it provides a simple way of preconditioning of the linear system resulting from (5.10).

Now let us return to the discussion of multilevel approximation. Having computed the cell average values $s_\omega(x_\omega)$, the interpolation on C_1 by Δs_1 at the first level, $\ell = 1$, in step **(4a)** of the adaptive multilevel Algorithm 20 is well-defined.

Adaptive Splitting of Cells at Finer Levels. Now let us turn to the construction of the subsequent point sets $C_{\ell+1}$, $\ell \geq 1$, of the finer levels in step **(4c)** of Algorithm 20. The construction of any point set $C_{\ell+1}$ depends on the approximation quality of the current representation s_ℓ of f at the previous level ℓ.

More precise, the approximation quality of s_ℓ is locally evaluated by computing the residual error

$$\eta_\omega = \max_{x \in X_\omega} |f(x) - s_\ell(x)| \tag{5.11}$$

for every *current* cell $\omega \in \mathcal{L}$. Whenever X_ω happens to be the empty set, we let $\eta_\omega = 0$. It is convenient to collect all current cells in $\mathcal{L}_\ell$.

Given the error indications $\{\eta_\omega\}_{\omega \in \mathcal{L}_\ell}$, further cells will be split, by using the function `split-cell` as follows. First, the cells are sorted according to their errors η_ω. Then, for a predetermined number $m \equiv m_\ell$, we split the m cells whose errors are largest. By these m splits, we obtain $m * 2^d$ new cells, and thus $m * 2^d$ new cell centers x_ω. The union $\mathcal{L}_{\ell+1} = \{\omega\}_{\omega \in \mathcal{L}}$ contains the current cells, so in particular $\mathcal{L}_{\ell+1}$ comprises all new cells.

We define the subsequent center set $C_{\ell+1}$ by the union

$$C_{\ell+1} = C_\ell \cup \{x_\omega : \omega \in \mathcal{L}_{\ell+1} \setminus \mathcal{L}_\ell \text{ and } X_\omega \text{ not empty }\}$$

of the previous center set C_ℓ and the centers of the new non-empty cells. So we obtain the next set $C_{\ell+1}$ in step **(4c)** of Algorithm 20, which comprises the previous one C_ℓ by construction, i.e., it satisfies $C_\ell \subset C_{\ell+1}$, as desired.

It remains to assign function values to the *new* centers in $C_{\ell+1} \setminus C_\ell$ (note that the centers in C_ℓ have values!). This is done in a similar way as for the initial center set C_1.

We first compute, for each new non-empty cell $\omega \in \mathcal{L}_{\ell+1} \setminus \mathcal{L}_\ell$, the polyharmonic spline interpolant s_ω of the form (3.44) satisfying $(f - s_\ell)\big|_{X_\omega} = s_\omega\big|_{X_\omega}$, before we assign the *cell average value* $s_\omega(x_\omega)$ to the cell center $x_\omega \in C_{\ell+1}$.

If, however, for any cell $\omega \in \mathcal{L}_{\ell+1}$, its corresponding point set X_ω fails to be $\mathcal{P}_k^d$-unisolvent, we assign the mean value

$$s_\omega(x_\omega) = \frac{1}{|X_\omega|} \sum_{x \in X_\omega} (f(x) - s_\ell(x))$$

to the cell center x_ω instead.

Having assigned a value to each $x_\omega \in C_{\ell+1} \setminus C_\ell$, the interpolation in step **(4a)** of Algorithm 20 is, at the next level $\ell+1$, well-defined. Indeed, note that the values for the centers at the previous set C_ℓ have already been computed at level ℓ. They are given by the values $s_\ell\big|_{C_\ell}$. So for the points in C_ℓ we have $(f - s_\ell)\big|_{C_\ell} = 0$ in step **(4a)** of Algorithm 20, when $\ell = \ell + 1$.

5.4 Hierarchical Surface Visualization

In this section, the performance of the different multilevel approximation schemes in this chapter is evaluated and compared. We considered involving three different methods. The first two methods, **SDF** and **AT1**, work with multilevel interpolation, Algorithm 19 of Section 5.2, but they use two different strategies for constructing the data hierarchy in (5.1). These two different strategies are scattered data filtering, **SDF** (see Section 4.4), and adaptive thinning, **AT1** (see Section 4.5). The other method is the adaptive multilevel approximation scheme, **AMA**, Algorithm 20 of Section 5.3.

5.4.1 Hurrungane — A Test Case from Terrain Modelling

In our numerical examples, we decided to work with one real-world data set from terrain modelling. This particular data set, called *Hurrungane*, is displayed in Figure 4.9 on page 100. The data is a sample of height values $\{f(x)\}_{x \in X}$ taken at $|X| = 23092$ distinct geographic locations of a Norwegian mountain area. The area is $\Omega = [437000, 442000] \times [6812000, 6817000] \subset \mathbb{R}^2$, where the samples were collected, and so $d = 2$ in this case. The minimum height of this data set is $\min_{x \in X} f(x) = 1100$ meters, and the maximum height is $\max_{x \in X} f(x) = 2400$ meters above sea-level.

5.4.2 Numerical Results and Comparisons

Now let us compare the performance of the adaptive multilevel approximation scheme in Section 5.3, **AMA**, against the other two, multilevel interpolation using scattered data filtering, **SDF**, and multilevel interpolation using the adaptive thinning algorithm **AT1** of Section 4.5. For the purpose of illustration, it is sufficient to work with merely two sublevels, i.e., $L = 2$ in (5.1) (when working with **SDF**, **AT1**), and in (5.7) (when working with **AMA**).

For the coarser set C_1, to be used in the scheme **AMA**, we selected the set which is displayed in Figure 5.1 **(b)**. This set comprises $n_1 = |C_1| = 840$ cell centers, whose corresponding cells are shown in Figure 5.1 **(a)**. In order to obtain the finer set C_2 in (5.7), we split $m = 419$ of these cells according to the magnitude of their cell error in (5.11). This yields $4 * 419 = 1676$ new cells, and thus 1676 new cell centers. But we found that 21 of these 1676 new cells were empty. Recall that the coarse set C_2 is the union of the points in C_1 and the centers of the non-empty new cells. Therefore, the size of C_2 is $n_2 = 840 + 1655 = 2495$. The two nested sets C_1 and C_2 are displayed in Figure 5.2 **(a)** and **(b)**.

For the purpose of comparing the method **AMA** with **SDF** and **AT1**, we have also generated two pairs of nested subsets, $X_1^F \subset X_2^F$, $X_1^A \subset X_2^A$, of equal sizes, i.e., $|X_1^F| = |X_1^A| = 840$ and $|X_2^F| = |X_2^A| = 2495$. The sets X_1^F, X_2^F, to be used by **SDF**, were generated by scattered data filtering (see Section 4.4), whereas the sets X_1^A, X_2^A were output by the adaptive thinning algorithm **AT1** (see Subsection 4.5.4). The four sets $X_1^F, X_2^F, X_1^A, X_2^A$ are shown in Figure 5.2 **(c)**-**(f)**.

Now the three different pairs of subsets $C_1 \subset C_2$, $X_1^F \subset X_2^F$, $X_1^A \subset X_2^A$ were used in order to compute three corresponding pairs of interpolants s_1, s_2 by Algorithm 19 (for **SDF**, **AT1**) and Algorithm 20 (for **AMA**). Recall that s_1 in (5.5) is the thin plate spline interpolant to the data at C_1 (when using the method **AMA**), X_1^F (when using **SDF**), and X_1^A (when using **AT1**). Moreover, $s_2 = s_1 + \Delta s_2$, where Δs_2, of the form (5.6), is the interpolant of the resulting residual error $f - s_1$ at C_2, X_2^F, or X_2^A. In either case, we selected $\varrho_2 = 250.0$ for the support radius of the compactly supported radial basis function ϕ_{ϱ_2} in (5.6), where we let $\phi(r) = (1-r)_+^4(4r+1)$, see Table 3.2 on page 34. For details on compactly supported radial basis functions, we refer to the discussion in Subsection 3.1.1 of Chapter 3, and the paper [170].

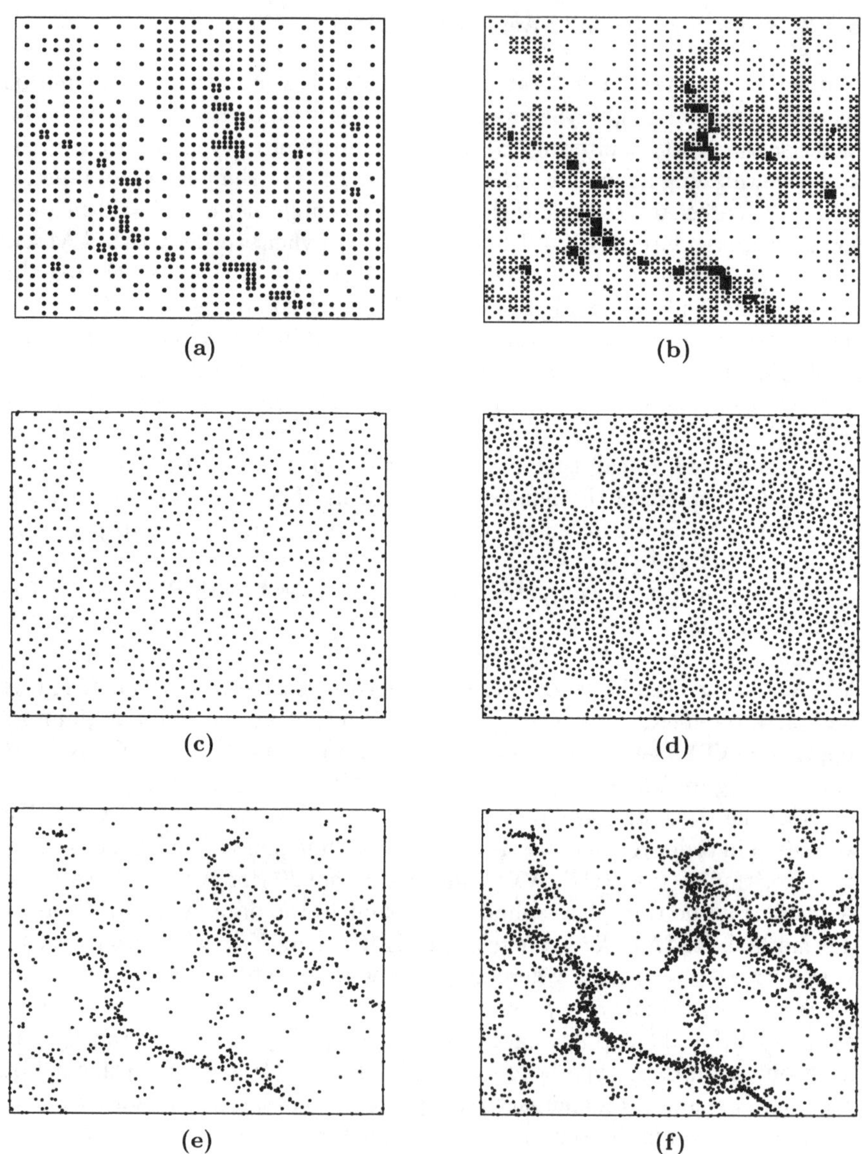

Fig. 5.2. Hurrungane. Subsets generated by **AMA**, **(a)** and **(b)**, scattered data filtering **SDF**, **(c)** and **(d)**, and adaptive thinning **AT1**, **(e)** and **(f)**.

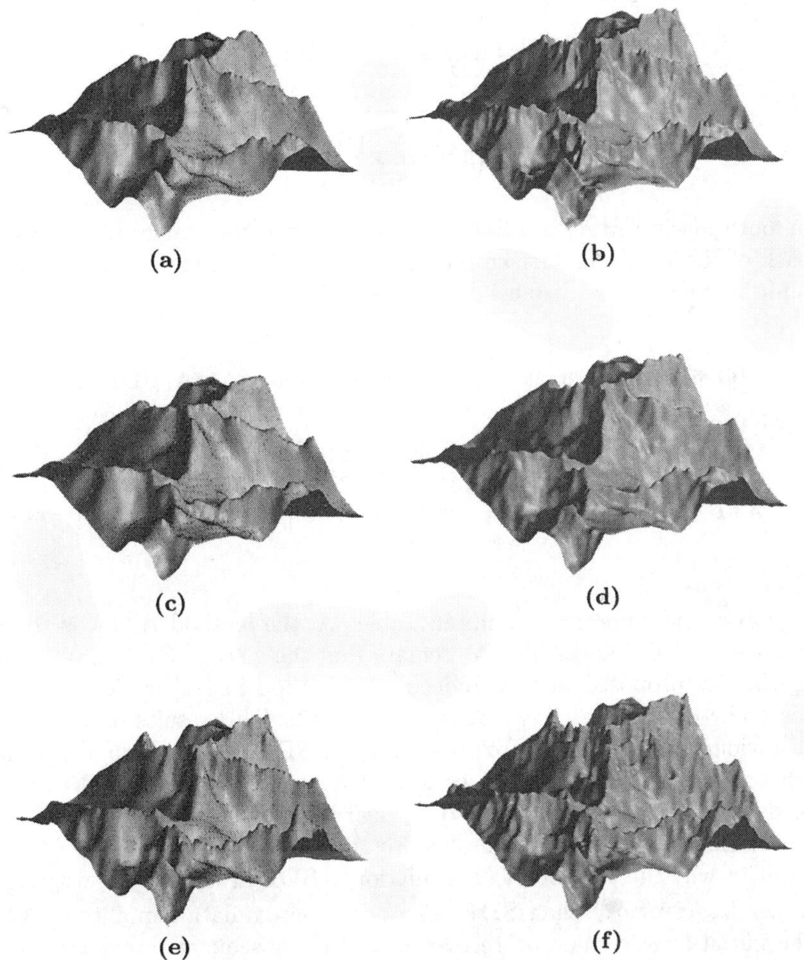

Fig. 5.3. Hurrungane. Surfaces of the coarse representation, s_1, and the fine representation, s_2, generated by the method **AMA**, **(a)** and **(b)**, the method **SDF**, **(c)** and **(d)**, and the method **AT1**, **(e)** and **(f)**.

The approximation quality of the three different methods **AMA**, **SDF**, and **AT1** was evaluated by recording the L_∞-error and the discrete L_2-error,

$$\eta_\infty^{(\ell)} = \max_{x \in X} |f(x) - s_\ell(x)|,$$

$$\eta_2^{(\ell)} = \sqrt{\frac{1}{|X|} \sum_{x \in X} |f(x) - s_\ell(x)|^2},$$

at both levels $\ell = 1, 2$. Table 5.1 reflects our numerical results. Moreover, each of the resulting surface representations of Hurrungane, the coarser s_1 and the finer s_2, are displayed in Figure 5.3.

Table 5.1. Approximation quality of the methods **AMA**, **SDF**, and **AT1**

Method	n_1	$\eta_\infty^{(1)}$	$\eta_2^{(1)}$	n_2	ϱ_2	$\eta_\infty^{(2)}$	$\eta_2^{(2)}$
AMA	840	152.404	21.957	2495	250.0	77.682	11.550
SDF	840	226.030	31.863	2495	250.0	134.069	16.164
AT1	840	341.381	33.998	2495	250.0	229.369	23.156

Given the numerical results in Table 5.1, the method **AMA** is the best, followed by **SDF** and **AT1**. We remark that the method **SDF** is very similar to the one proposed in [72]. Indeed, the method in [72] works with a data hierarchy (5.1) of *uniformly distributed* subsets. This results in a fairly good behaviour of the initial approximation s_1 for **SDF**, see Figure 5.3 **(c)**. In contrast to this, the method **AT1** works with unevenly distributed subsets, which leads to undesirable overshoots of s_1 near the clusters in X_1^A (corresponding to the ridges of the mountains, see Figure 5.2 **(e)** and Figure 5.3 **(e)**). This explains why the method **AT1** is inferior to **SDF** at level $\ell = 1$, with a much larger L_∞-error $\eta_\infty^{(1)}$ than **SDF**. The poor approximation quality of **AT1** at the initial level cannot be recovered by the subsequent interpolant s_2, see Figure 5.3 **(f)**. Indeed, the method **AT1** continues to be inferior to **SDF** at level $\ell = 2$. This is also supported by the numerical results in [96].

In conclusion, the performance of multilevel interpolation relies heavily on the data hierarchy (5.1). Moreover, as also shown in [96], the approximation quality of the initial interpolant s_1 has a strong effect on the approximation quality of subsequent interpolants. Now note that the approximation quality of the method **AMA** is, when compared with **SDF** and **AT1**, much better at the initial level. Indeed, the method **AMA** reduces the approximation errors $\eta_\infty^{(1)}$ and $\eta_2^{(1)}$ of the method **SDF** by approximately a third. This is due to the well-balanced distribution of the points in the coarse set C_1, see Figure 5.2 **(a)**. The distribution of the points in C_1 is not as clustered as in the set X_1^A, see Figure 5.2 **(e)**. This helps to avoid the abovementioned overshoots of the initial interpolant s_1, see Figure 5.3 **(a)**. The method **AMA** continues to be superior to both **SDF** and **AT1** at the coarse level $\ell = 2$. This complies with our above explanation concerning the corresponding comparison between **SDF** and **AT1**.

In summary, the new adaptive multilevel approximation scheme of Section 5.3 yields a good alternative to the multilevel interpolation scheme of Section 5.2. The good performance of the method **AMA** is mainly due to the sophisticated construction of the subsets C_1 and C_2. On the one hand, unlike **SDF**, this construction is *data-dependent*. On the other hand, in contrast to **AT1**, dense clusters are widely avoided, and this helps to damp down possible overshoots of the interpolants s_ℓ, $\ell = 1, 2$.

6 Meshfree Methods for Transport Equations

Meshfree methods are recent and modern discretization techniques for numerically solving partial differential equations (PDEs). In contrast to the well-established traditional methods, such as *finite differences* (FD), *finite volumes* (FV), and *finite element methods* (FEM), meshfree methods do not require sophisticated algorithms and data structures for maintaining a grid, which is often the most time consuming task in mesh-based simulations.

Moreover, meshfree methods provide flexible, robust and reliable discretizations, which are particularly suited for multiscale simulation. Consequently, meshfree discretizations have recently gained much attention in many different applications from computational sciences and engineering, as well as in numerical analysis. Among a few others, the currently most prominent meshfree discretization techniques are *smoothed particle hydrodynamics* (SPH) [123], the *partition of unity method* (PUM) [5, 120], and *radial basis functions* (RBF), see Chapter 3.

In this chapter, a novel adaptive meshfree method of (backward) characteristics, **AMMoC**, for multiscale simulation of tranport pocesses is proposed. The method **AMMoC** combines an adaptive Lagrangian particle method with local scattered data interpolation by polyharmonic splines. The adaption strategy is built on customized rules for the refinement and coarsening of current nodes, each at a time corresponding to one flow particle. The required adaption rules are constructed by using available results concerning local error estimates and numerical stability of polyharmonic spline interpolation, as discussed in Section 3.8 of Chapter 3.

The outline of this chapter is as follows. In the following Section 6.1, a short discussion on transport equations is provided, before the basic ingredients of our method **AMMoC** are explained in Section 6.2. The construction of adaption rules is discussed in Section 6.3. Finally, Section 6.4 is devoted to numerical simulation of various multiscale phenomena in flow modelling.

The model problems discussed in Section 6.4 comprise tracer transportation over the artic, Burgers equation, and two-phase fluid flow in porous medium. The latter is done by using the *five-spot problem*, a popular model problem from hydrocarbon reservoir simulation, where **AMMoC** is shown to be competitive with two leading commercial reservoir simulators, ECLIPSE and FrontSim of Schlumberger.

6.1 Transport Equations

Many physical phenomena in transport processes are described by time-dependent *hyperbolic conservation laws.* Their governing equations have the form

$$\frac{\partial u}{\partial t} + \nabla f(u) = 0 \tag{6.1}$$

where for some domain $\Omega \subset \mathbb{R}^d$, $d \geq 1$, and a compact time interval $I = [0, T]$, $T > 0$, the unknown function $u : I \times \Omega \to \mathbb{R}$ corresponds to a physical quantity, such as saturation or concentration density. Moreover, the function $f(u) = (f_1(u), \ldots, f_d(u))^T$ in (6.1) denotes the *flux tensor.*

In this chapter, we consider numerically solving (6.1) on given initial conditions

$$u(0, x) = u_0(x), \qquad \text{for } x \in \Omega = \mathbb{R}^d, \tag{6.2}$$

where we assume that the solution u of the resulting initial value problem has compact support in $I \times \mathbb{R}^d$. The latter serves to avoid considering explicit boundary conditions.

In situations where the flux tensor f is a *linear* function, i.e.,

$$f(u) \equiv \mathbf{v} \cdot u, \tag{6.3}$$

we obtain the *linear* advection equation

$$\frac{\partial u}{\partial t} + \mathbf{v} \cdot \nabla u = 0 \tag{6.4}$$

provided that the given *velocity field*

$$\mathbf{v} = \mathbf{v}(t, x), \qquad t \in I, x \in \Omega,$$

is divergence-free, i.e.,

$$\operatorname{div}\mathbf{v} = \sum_{j=1}^{d} \frac{\partial \mathbf{v}_j}{\partial x_j} \equiv 0.$$

This special case of (6.1) is also often referred to as *passive advection.* In this case, the scalar solution u is constant along the *streamlines* of (flow) particles, and the shapes of these streamlines are entirely and uniquely determined by the velocity field $\mathbf{v}$.

In previous work [10], an adaptive meshfree method for solving linear advection equations of the above form is proposed. The method in [10] is a combination of an adaptive version of the semi-Lagrangian method (**SLM**) and the meshfree radial basis function interpolation (RBF). The resulting advection scheme is then in [11] used for the simulation of tracer transportation in the arctic stratosphere. Selected details on this challenging model problem are explained later in this chapter, see Subsection 6.4.1.

The semi-Lagrangian method, used in [10], traces the path of flow particles along the trajectories of their streamlines. Therefore, the scheme in [10] is essentially a *method of characteristics* (**MoC**), see [34, 83] for an overview. Indeed, the characteristic curves of the equation (6.4) coincide with the streamlines of flow particles, and the meshfree **SLM** in [10, 11] captures the flow of a discrete set of particles along their characteristic curves. This is accomplished by computing backward trajectories for a finite set of current particles (nodes) at each time step, whereas the node set is adaptively modified during the simulation. The applications of the scheme in [10] are, however, restricted to *linear* transport problems with governing equations of the form (6.4).

In this chapter, the advection scheme of [10] is extended in the following section, so that an adaptive meshfree method of (backward) characteristics, called **AMMoC**, is obtained, which also covers the numerical treatment of *nonlinear* equations of the form (6.1). To this end, we follow along the lines of our previous work [12]. We remark that the nonlinear case is much more complicated than the linear one, and thus any possible generalization of the **SLM** in [10] requires care. Indeed, in contrast to the linear case, a nonlinear flux function f usually leads to *discontinuities* in the solution u, *shocks*, as observed in many relevant applications, such as fluid and gas dynamics. In such situations, the classical **MoC** becomes unwieldy or impossible, as the evolution of the flow along the characteristic curves is typically very complicated, or characteristic curves may even be undefined (see [49, Subsection 6.3.1] for a discussion on these and related phenomena).

Now in order to be able to model the behaviour of the solution with respect to shock formation and shock propagation we work with a *vanishing viscosity* approach, yielding the modified advection-diffusion equation

$$\frac{\partial u}{\partial t} + \nabla f(u) = \epsilon \cdot \Delta u, \tag{6.5}$$

where the parameter $\epsilon > 0$ is referred to as the *diffusion coefficient.* In this way, the solution u of the *hyperbolic* equation (6.1) is approximated arbitrarily well by the solution of the modified *parabolic* equation (6.5), provided that the parameter ϵ is sufficiently small. This modification is a standard stabilization technique for nonlinear equations, dating back to Burgers [25], who utilized a flux function of the form

$$f(u) = \frac{1}{2}u^2 \cdot r, \tag{6.6}$$

with some flow direction $r \in \mathbb{R}^d$, for modelling free turbulences in fluid dynamics. The resulting *Burgers equation* is nowadays a popular standard test case for nonlinear transport equations. We come back to this test case in Subsection 6.4.2.

Later, in Subsection 6.4.3, we consider solving the *Buckley-Leverett equation*, whose *non-convex* flux function has the form

$$f(u) = \frac{u^2}{u^2 + \mu(1-u)^2} \cdot r\,. \tag{6.7}$$

The Buckley-Leverett equation models the saturation of a two-phase flow in a porous medium with neglecting gravitational forces or capillary effects. In this case, the value of μ in (6.7) is the ratio of the two different fluid's viscosities. This model problem is typically encountered in *oil reservoir modelling*. Details are explained in Subsection 6.4.3.

6.2 Meshfree Method of Characteristics

Let us first explain the semi-Lagrangian method (**SLM**), as used in [10], for solving the passive advection equation (6.4), before we discuss its generalization, the method of (backward) characteristics (**MoC**), later in this section. For a more comprehensive discussion on the **SLM** and its applications in meteorology, we refer to the survey [165] and the textbook [124, Section 7].

The **SLM** integrates the Lagrangian form

$$\frac{du}{dt} = 0 \tag{6.8}$$

of the linear advection equation (6.4) along trajectories. Therefore, the starting point of the **SLM** is the discretization

$$\frac{u(t+\tau,\xi) - u(t,x^-)}{\tau} = 0, \qquad \xi \in \Omega,$$

of (6.8), where $t \in I$ is the current time, $\tau > 0$ the time step size, and $x^- \equiv x^-(\xi)$ denotes the *upstream point* of any ξ. The point x^- is the unique location of that particle at time t, which by traversing along its trajectory arrives at ξ at time $t+\tau$, see Figure 6.1. In particular, $u(t+\tau,\xi) = u(t,x^-)$.

In order to numerically solve (6.4), the meshfree advection scheme in [10] works with a finite discrete set $\Xi \subset \Omega$ of *current* nodes, each of whose elements $\xi \in \Xi$ corresponds, at time t, to one flow particle. At each time step $t \to t+\tau$, new values $u(t+\tau,\xi)$, $\xi \in \Xi$, are computed from the *current* values $u(t,\xi)$, $\xi \in \Xi$. This is done by using the following semi-Lagrangian advection scheme, Algorithm 21. Initially, at time $t = 0$, the nodes in Ξ are randomly distributed in Ω, and the values $u(0,\xi)$, $\xi \in \Xi$, are given by the initial condition (6.2).

We remark that computing the upstream point x^- of any node $\xi \in \Xi$ amounts to solving the *ordinary differential equation* (ODE)

$$\dot{x} = \frac{dx}{dt} = \mathbf{v}(t,x) \tag{6.9}$$

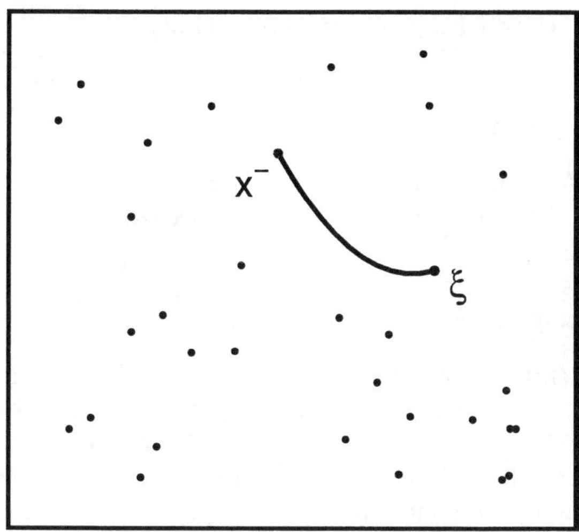

Fig. 6.1. The point $x^- = \Phi^{t,t+\tau}\xi$ is the upstream point of the node $\xi \in \Xi$.

on initial condition $x(t+\tau) = \xi$, so that $x(t) = x^-$.

Adopting some standard notation from dynamic systems, we can express the upstream point x^- of ξ as

$$x^- = \Phi^{t,t+\tau}\xi, \tag{6.10}$$

where $\Phi^{t,t+\tau} : \Omega \to \Omega$ denotes the *continuous evolution* of the (backward) flow of (6.9). An equivalent formulation for (6.10) is given by $\xi = \Phi^{t+\tau,t}x^-$, since $\Phi^{t+\tau,t}$ is the inverse of $\Phi^{t,t+\tau}$.

Now since the solution u of (6.4) is constant along the trajectories of the flow particles, we have $u(t,x^-) = u(t+\tau,\xi)$, and so the desired values $u(t+\tau,\xi)$, $\xi \in \Xi$, may immediately be obtained from the upstream point values $u(t,x^-)$. But in general, neither the *exact* location of x^-, nor the value $u(t,x^-)$ is known.

Therefore, during the performance of the flow simulation, this requires first computing an approximation $\tilde{x}$ to the upstream point $x^- = \Phi^{t,t+\tau}\xi$ for each $\xi \in \Xi$. It is convenient to express the approximation $\tilde{x}$ to x^- as

$$\tilde{x} = \Psi^{t,t+\tau}\xi,$$

where $\Psi^{t,t+\tau} : \Omega \to \Omega$ is the *discrete evolution* of the flow, corresponding to the continuous evolution $\Phi^{t,t+\tau}$ in (6.10) [44]. The operator $\Psi^{t,t+\tau}$ is given by any specific numerical method for solving the above ODE (6.9).

Having computed $\tilde{x}$, the value $u(t,\tilde{x})$ is then determined from the current values $\{u(t,\xi)\}_{\xi\in\Xi}$ by local interpolation. Altogether, the above discussion leads us to the following algorithm concerning the advection step $t \to t+\tau$ of the semi-Lagrangian method.

Algorithm 21 (Semi-Lagrangian Advection).

INPUT: *Node set* Ξ, *values* $\{u(t,\xi)\}_{\xi\in\Xi}$, *and time step size* τ;

FOR *each* $\xi \in \Xi$ **DO**

(1) *Compute the upstream point approximation* $\tilde{x} = \Psi^{t,t+\tau}\xi$;
(2) *Determine the value* $u(t,\tilde{x})$ *by local interpolation;*
(3) *Advect by letting* $u(t+\tau,\xi) = u(t,\tilde{x})$;

OUTPUT: *Values* $u(t+\tau,\xi)$ *for all* $\xi \in \Xi$.

The implementation of the two steps, **(1)** and **(2)**, in Algorithm 21, deserves some further comments. Let us first turn to step **(1)**. Recall that this requires solving, for any current node $\xi \in \Xi$, the initial value problem (6.9). This end end, we follow along the lines of the seminal paper of Robert [142] (see also [165] and [124, equation (7.66a)]), where it is recommended to employ a recursion of the form

$$\beta_{k+1} = \tau \cdot \mathbf{v}(t+\tau/2, \xi - \beta_k/2) \tag{6.11}$$

in order to obtain after merely a few iterations a sufficiently accurate linear approximation $\beta \in \mathbb{R}^d$ of the trajectory arriving at ξ, and so then let

$$\tilde{x} = \xi - \beta,$$

see Figure 6.2.

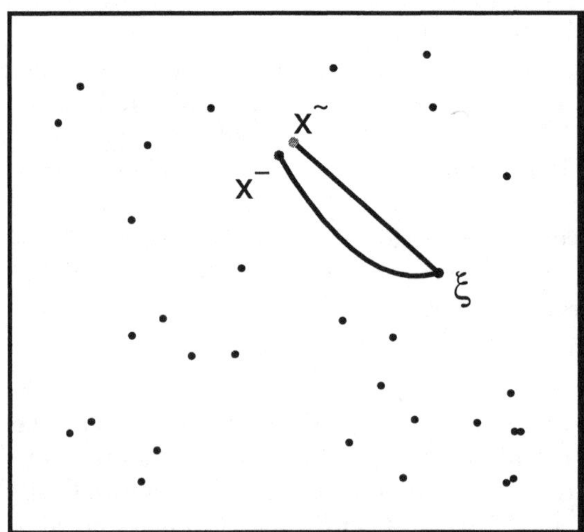

Fig. 6.2. The point $\tilde{x} = \Psi^{t,t+\tau}\xi$ is an approximation to $x^- = \Phi^{t,t+\tau}\xi$, being the upstream point of $\xi \in \Xi$.

As regards step **(2)**, the local interpolation around any upstream point approximation $\tilde{x}$ is done by using polyharmonic splines, see Section 3.8. To be more precise, the values of $u(t,\cdot)$ at a set $\mathcal{N} \equiv \mathcal{N}(\tilde{x}) \subset \Xi$ of current nodes in the neighbourhood of $\tilde{x}$ are used in order to solve the interpolation problem

$$s(\nu) = u(t,\nu), \qquad \text{for all } \nu \in \mathcal{N},$$

where s is the polyharmonic spline interpolant of the form (3.44). The approximation of $u(t,\tilde{x})$, required in step **(2)**, is then obtained by $s(\tilde{x}) \approx u(t,\tilde{x})$.

We remark that semi-Lagrangian advection, Algorithm 21, is unconditionally stable, provided that the local interpolation is stable. This is in contrast to Eulerian schemes, which, for the sake of stability, typically work with very small time steps [113]. For a concise analysis concerning the convergence and stability of semi-Lagrangian methods, we refer to the paper [62] by Falcone and Ferretti.

Now let us turn to the *nonlinear* case (where the flux function f is nonlinear). In this case, we are concerned with solving the advection-diffusion equation (6.5). Similar to the linear case, this leads us to the discretization

$$\frac{u(t+\tau,\xi) - u(t,x^-)}{\tau} = \epsilon \Delta u(t,x^-) \tag{6.12}$$

of its corresponding Lagrangian form

$$\frac{du}{dt} = \epsilon \cdot \Delta u,$$

where $\frac{du}{dt} = \frac{\partial u}{\partial t} + \nabla f(u)$ is the *material derivative.*

As in the linear case, we integrate the Lagrangian form of the equation (6.5) along streamlines of flow particles, characteristic curves. Therefore, having computed, for any current node $\xi \in \Xi$, an approximation $\tilde{x} = \Psi^{t,t+\tau}\xi$ to the upstream point $x^- = \Phi^{t,t+\tau}\xi$, the desired approximation to $u(t+\tau,\xi)$ is then given by

$$u(t+\tau,\xi) = u(t,\tilde{x}) + \tau \cdot \epsilon \Delta u(t,\tilde{x}), \qquad \text{for } \xi \in \Xi. \tag{6.13}$$

However, note that in contrast to passive advection, the characteristic curves of the equation (6.5) depend also on u. In particular, the advection velocity $\mathbf{v} = \frac{\partial f(u)}{\partial u}$ depends on u. In order to compute the velocity

$$\mathbf{v}\,(t+\tau/2, \xi - \beta_k/2)$$

at intermediate time $t+\tau/2$, as required in the iteration (6.11), we employ the linear extrapolation

$$\mathbf{v}\left(t+\frac{\tau}{2},\cdot\right) = \frac{3}{2}\mathbf{v}(t,\cdot) - \frac{1}{2}\mathbf{v}(t-\tau,\cdot), \tag{6.14}$$

where, again, we use polyharmonic spline interpolation for computing the values of $\mathbf{v}(t,\cdot), \mathbf{v}(t-\tau,\cdot)$ on the right hand side in (6.14) from the available values $\{\mathbf{v}(t,\xi)\}_{\xi\in\Xi}$ and $\{\mathbf{v}(t-\tau,\xi)\}_{\xi\in\Xi}$ at the nodes. Initially, for obtaining the required values of $u(\tau,\cdot)$ from the given initial conditions (6.2), we use a generalized *two-level* Lax-Friedrich scheme on $\Xi^0 \equiv \Xi^\tau$.

Altogether, this leads us to the following algorithm, method of (backward) characteristics (**MoC**), for the advection step $t \to t+\tau$ in the nonlinear case.

Algorithm 22 (Method of Characteristics).

INPUT: *Nodes* Ξ, *values* $\{u(t,\xi)\}_{\xi\in\Xi}$, *time step* τ, *diffusion coefficient* ϵ;

FOR *each* $\xi \in \Xi$ **DO**

(1) *Compute the upstream point approximation* $\tilde{x} = \Psi^{t,t+\tau}\xi$;
(2) *Determine the values* $u(t,\tilde{x})$ *and* $\Delta u(t,\tilde{x})$ *by local interpolation;*
(3) *Advect by letting* $u(t+\tau,\xi) = u(t,\tilde{x}) + \tau\cdot\epsilon\Delta u(t,\tilde{x})$;

OUTPUT: *The values* $u(t+\tau,\xi)$ *for all* $\xi\in\Xi$, *at time* $t+\tau$.

Step **(2)** of Algorithm 22 deserves a comment concerning the interpolation of the value $\Delta u(t,\tilde{x})$. Similar as in Algorithm 21, we work with local interpolation by polyharmonic splines, but with a smoother basis function. For instance, note that in two dimensions the Laplacian of the thin plate spline $\phi_{2,2}(r) = r^2\log(r)$ (and thus the Laplacian of the resulting interpolant) has a singularity at zero. In this case, we prefer to work with the *smoother* basis function $\phi_{2,3}(r) = r^4\log(r)$, whose Laplacian $\Delta\phi_{2,3}(\|x\|)$ is well-defined everywhere. Further details on this are explained in [12].

6.3 Adaption Rules

Having computed the values $u(t+\tau,\xi)$, for all $\xi\in\Xi$, via (6.13), the current node set $\Xi\equiv\Xi^t$ (at time t) is modified by the removal (coarsening), and the insertion (refinement) of nodes, yielding a new node set $\Xi\equiv\Xi^{t+\tau}$ (at time $t+\tau$). The adaption of the nodes relies on a customized a posteriori error indicator, to be explained in the following of this section, see also [11].

Error Indication. An effective strategy for the adaptive modification of the nodes requires well-motivated refinement and coarsening rules as well as a customized error indicator. We understand the error indicator $\eta:\Xi\to[0,\infty)$ as a function of the current node set $\Xi\equiv\Xi^t$ (at time t) which assigns a *significance* value $\eta(\xi)$ to each node $\xi\in\Xi$. The value $\eta(\xi)$ is required to reflect the local approximation quality of the interpolation around $\xi\in\Xi$. The significances $\eta(\xi)$, $\xi\in\Xi$, are then used in order to flag single nodes $\xi\in\Xi$ as *to be refined* or *to be coarsened* according to the following criteria.

Definition 23. *Let* $\eta^* = \max_{\xi \in \Xi} \eta(\xi)$, *and* $\theta_{\text{crs}}, \theta_{\text{ref}}$ *be two threshold values satisfying* $0 < \theta_{\text{crs}} < \theta_{\text{ref}} < 1$. *We say that a node* $\xi \in \Xi$ *is* `to be refined`, *iff* $\eta(\xi) > \theta_{\text{ref}} \cdot \eta^*$, *and* ξ *is* `to be coarsened`, *iff* $\eta(\xi) < \theta_{\text{crs}} \cdot \eta^*$.

Note that a node ξ cannot be refined and be coarsened at the same time; in fact, it may neither be refined nor be coarsened.

Now let us turn to the definition of the error indicator η. To this end, we follow along the lines of our previous paper [84], where a scheme for the detection of discontinuities of a surface, *fault lines*, from scattered data was developed. We let

$$\eta(\xi) = |u(\xi) - s(\xi)|,$$

where $s \equiv s_{\mathcal{N}}$ denotes the polyharmonic spline interpolant, which matches the values of $u \equiv u(t, \cdot)$ at a neighbouring set $\mathcal{N} \equiv \mathcal{N}(\xi) \subset \Xi \setminus \xi$ of current nodes, i.e., $s(\nu) = u(\nu)$ for all $\nu \in \mathcal{N}$. In our numerical examples for bivariate data, where $d = 2$, we work with local thin plate spline interpolation. Recall that this particular interpolation scheme reconstructs linear polynomials. In this case, the value $\eta(\xi)$ vanishes whenever u is linear around ξ. Moreover, the indicator $\eta(\xi)$ is small whenever the *local* reproduction quality of the interpolant s is good. In contrast to this, a high value of $\eta(\xi)$ typically indicates that u is subject to strong variation locally around ξ.

Coarsening and Refinement. In order to balance the approximation quality of the model against the required computational complexity we insert new nodes into regions where the value of η is high (refinement), whereas we remove nodes from Ξ in regions where the value of η is small (coarsening).

To avoid additional computational overhead and complicated data structures, effective adaption rules are required to be as simple as possible. In particular, these rules ought to be given by *local* operations on the current node set Ξ. The following coarsening rule is in fact very simple and, in combination with the refinement, it turned out to be very effective as well.

Coarsening. A node $\xi \in \Xi$ is *coarsened* by its removal from the current node set Ξ, i.e., Ξ is modified by replacing Ξ with $\Xi \setminus \xi$.

As to the refinement rules, these are constructed on the basis of available local error estimates for polyharmonic spline interpolation. Recall from Subsection 3.8.2 that for any $x \in \Omega$, the pointwise error $|u(x) - s(x)|$ between (a sufficiently smooth) function $u \equiv u(t, \cdot)$ and the polyharmonic spline interpolant $s \equiv s_{\mathcal{N}}$ satisfying $s\big|_{\mathcal{N}} = u\big|_{\mathcal{N}}$ can be bounded above by an estimate of the form

$$|u(x) - s(x)| \leq C \cdot h_{\mathcal{N},\varrho}^{k-d/2}(x), \tag{6.15}$$

where $C > 0$ is a constant depending on u, and (for some radius $\varrho > 0$)

$$h_{\mathcal{N},\varrho}(x) = \sup_{\|y-x\|<\varrho} d_{\mathcal{N}}(y)$$

is the local *fill distance* of $\mathcal{N}$ around x.

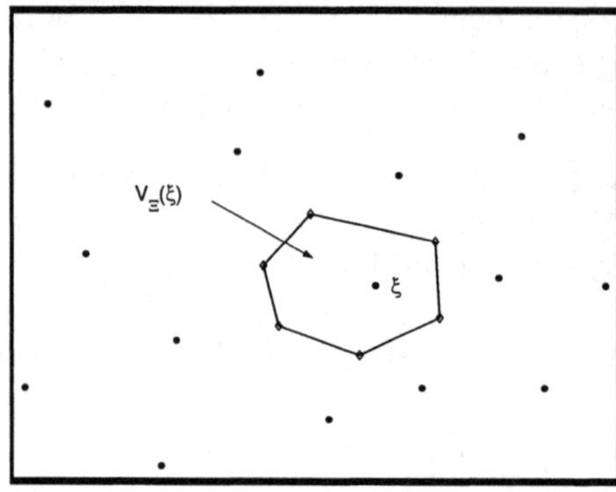

Fig. 6.3. Refinement of the node ξ. The Voronoi points ($\diamond$) are inserted.

Now, in the refinement of any node $\xi \in \Xi$, the desired reduction of the local error (6.15) around ξ is accomplished by reducing the distance function $d_{\mathcal{N}} = \min_{\nu \in \mathcal{N}} \| \cdot - \nu \|$ in a local neighbourhood of ξ. This in turn is done as follows. Recall the discussion on Voronoi diagrams in Section 2.3. The Voronoi tile

$$V_{\Xi}(\xi) = \left\{ y \in \mathbb{R}^d \,:\, d_{\Xi}(y) = \|y - \xi\| \right\} \subset \mathbb{R}^d$$

of ξ w.r.t. the point set Ξ is a convex polyhedron whose vertices, called *Voronoi points*, form a finite point set $\mathcal{V}_{\xi}$ in the neighbourhood of ξ. Figure 6.3 shows the Voronoi tile $V_{\Xi}(\xi)$ of a node ξ along with the set $\mathcal{V}_{\xi}$ of its Voronoi points.

Now observe that for any $\xi \in \mathcal{N}$, the distance function $d_{\mathcal{N}}$ is convex on $V_{\Xi}(\xi)$. Moreover, the function $d_{\mathcal{N}}$ has local maxima at the Voronoi points in $\mathcal{V}_{\xi}$. Altogether, this gives rise to define the local refinement of nodes as follows.

Refinement. A node $\xi \in \Xi$ is *refined* by the insertion of its Voronoi points into the current node set Ξ, i.e., Ξ is modified by replacing Ξ with $\Xi \cup \mathcal{V}_{\xi}$.

6.4 Multiscale Flow Simulation

The combination of either the semi-Lagrangian advection, Algorithm 21, or the method of (backward) characteristics, Algorithm 22, with both the meshfree local polyharmonic spline interpolation and the customized adaption rules, as developed in the previous Section 6.3, provides a flexible, robust and reliable discretization scheme for both linear and nonlinear transport equations. We refer to the resulting advection scheme as **A**daptive **M**eshfree **M**ethod **o**f **C**haracteristics, in short **AMMoC**. The good performance of **AMMoC** is in this section shown by using the following model problems, all of which are essentially requiring multiscale simulation.

The test case of the following Subsection 6.4.1, concerning passive advection, is also considered in our previous work [11], and the Diploma thesis [132] of Pöhn, where special emphasis is placed on implementation. In this test case, **AMMoC** combines Algorithm 21 with the adaption rules of Section 6.3. Then, in the remaining two subsections, *nonlinear* test cases are considered. In Subsection 6.4.2, we consider solving Burgers equation, and Subsection 6.4.3 concerns the modelling of two-phase flow in porous media, see also [12] and the related PhD thesis [106] of Käser. In this case, **AMMoC** combines Algorithm 22 with the adaption rules of Section 6.3.

6.4.1 Tracer Advection over the Arctic

This first test case scenario is concerning the simulation of tracer transportation in the arctic stratosphere. Let us briefly provide some background information on this challenging application. When investigating ozone depletion over the arctic, one interesting question is whether air masses with low ozone concentration are advected into southern regions. For the purpose of illustration, we merely work with a *simplified* advection model. Simplified, mainly because no realistic initial tracer concentration is assumed, and moreover no chemical reactions are included in the model. Nevertheless, we work with a realistic wind field, leading to quite typical filamentations of the tracer cloud, as corresponding to previous airborne observations [19].

The wind data, giving the velocity field $\mathbf{v}$ in (6.4), were taken from the *high-resolution regional climate model* (HIRHAM) [43]. HIRHAM resolves the arctic region with a horizontal resolution of $0.5°$. It is forced at the lateral and lower boundaries by ECMWF reanalysis data. We consider the transport of a passive tracer at 73.4 hPa in the vortex. This corresponds to an altitude of 18 km. The wind field reproduces the situation in January 1990.

Since stratospheric motion is thought to be constrained largely within horizontal layers, we use a two-dimensional horizontal transport scheme here. The wind data is represented in the corresponding two-dimensional coordinates of the horizontal layers in the three-dimensional HIRHAM model. The wind field $\mathbf{v}$ at initial time is shown in Figure 6.4 **(a)**, and the initial tracer distribution u_0 for the advection experiment is shown in Figure 6.4 **(b)**.

We remark that this application originated from the investigations in [9], where the formation of small-scale filaments is shown by using an adaptive tracer transport model. The simulation in [9] works with triangular meshes, thus it is mesh-based. For more details on this particular mesh-based method, including aspects concerning its parallelization, we refer to the paper [8] by Behrens.

Figure 6.5 shows a comparison between our *meshfree* method **AMMoC** and the *mesh-based* adaptive **SLM** [8] of Behrens, each recorded at day 15 of the simulation. Note that both simulations capture the fine filamentation of the tracer very well. The corresponding node distribution of the meshfree simulation by **AMMoC** is shown in Figure 6.6.

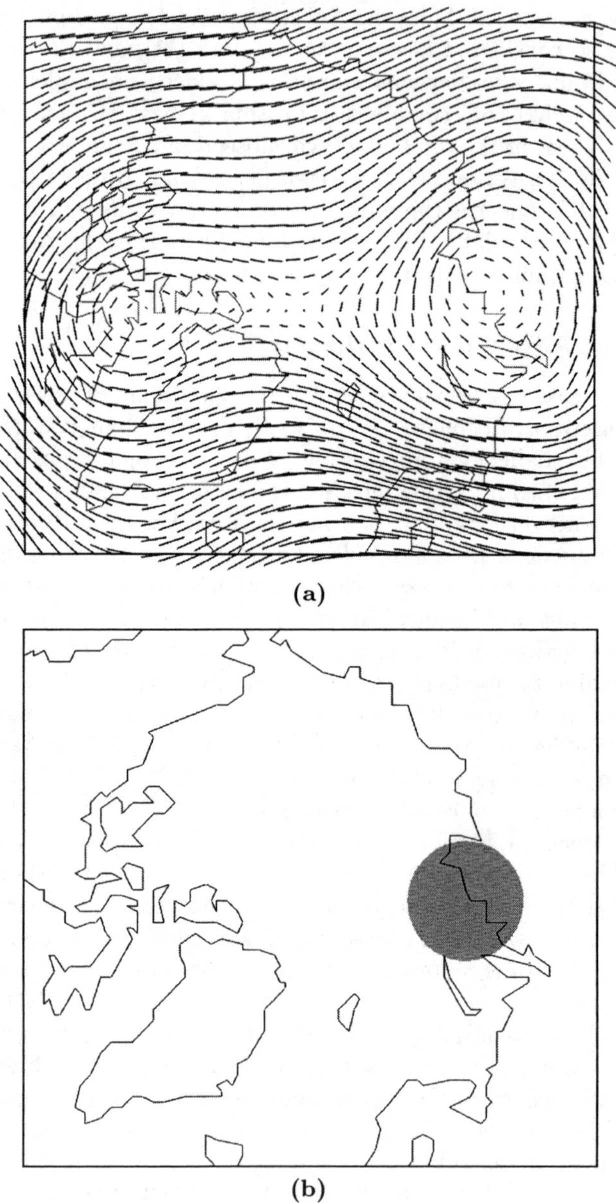

Fig. 6.4. Tracer advection over the artic. **(a)** Wind field $\mathbf{v}$ at initial time; **(b)** initial tracer distribution u_0. Continental outlines are given for orientation (Greenland in the lower left part).

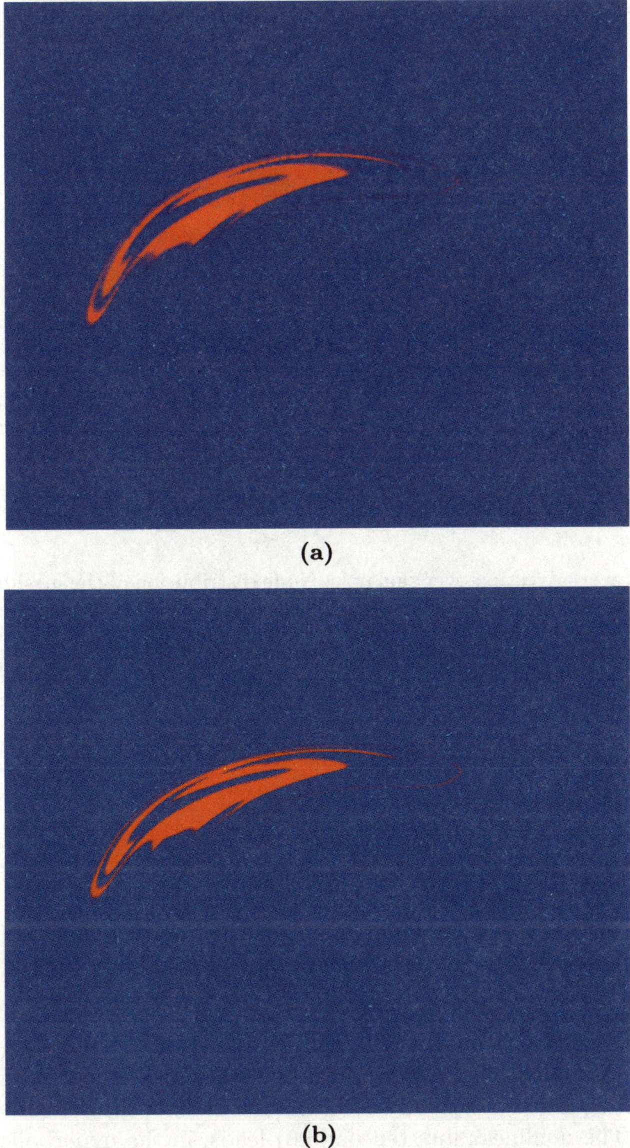

Fig. 6.5. Tracer advection over the artic. Comparison between **(a)** the mesh-based method [8] and **(b)** our meshfree method **AMMoC** by using two snapshots, each recorded at day 15 of the simulation.

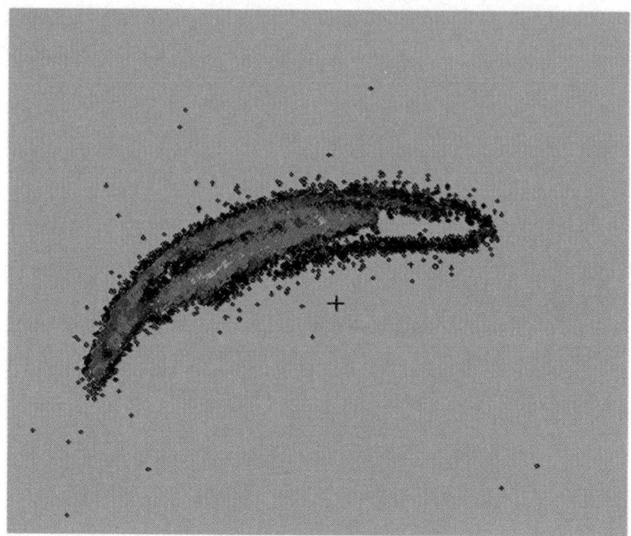

Fig. 6.6. Tracer advection over the artic. Node distribution of the meshfree simulation by **AMMoC** at day 15. Pink colour indicates nodes with tracer concentration above 0.2.

6.4.2 Burgers Equation — A Nonlinear Standard Test

The viscous Burgers equation

$$\frac{\partial u}{\partial t} + u\nabla u \cdot r = \epsilon \cdot \Delta u, \tag{6.16}$$

introduced in [25] as a mathematical model of free turbulence in fluid dynamics, is nowadays a popular standard test case, mainly for the following two reasons.

(1) Burgers equation contains the simplest form of a nonlinear advection term $u \cdot \nabla u$ simulating the physical phenomenon of wave motion;

(2) Burgers equation is typically used as a test case for the modelling of *shock waves*. The nonlinear flux tensor (6.6) leads, in the hyperbolic equation (6.1), to *shocks*. As soon as the shock front occurs, there is no classical solution of the equation (6.1), and its weak solution becomes discontinuous. The modified parabolic equation (6.16) has for $t > 0$ a *smooth* solution u_ϵ which approximates (for sufficiently small ϵ) the shock front propagation arbitrarily well.

We use Burgers equation (6.16) as a preliminary test case for our adaptive meshfree advection scheme **AMMoC**, before we turn to a more difficult nonlinear transport problem in the following subsection.

In this subsection, we consider solving Burgers equation (6.16) in combination with the compactly supported initial condition

$$u_0(x) = \begin{cases} \exp\left(\frac{\|x-c\|^2}{\|x-c\|^2-R^2}\right) & \text{for } \|x-c\| < R, \\ 0 & \text{otherwise,} \end{cases}$$

where $R = 0.25$, $c = (-0.25, -0.25)$, and we let the unit square $\Omega = [0,1]^2$ be the computational domain. This test case is also used in [77]. Figure 6.7 shows the initial condition, and the flow field $r = (1,1)^T$, yielding a flow along the diagonal in Ω.

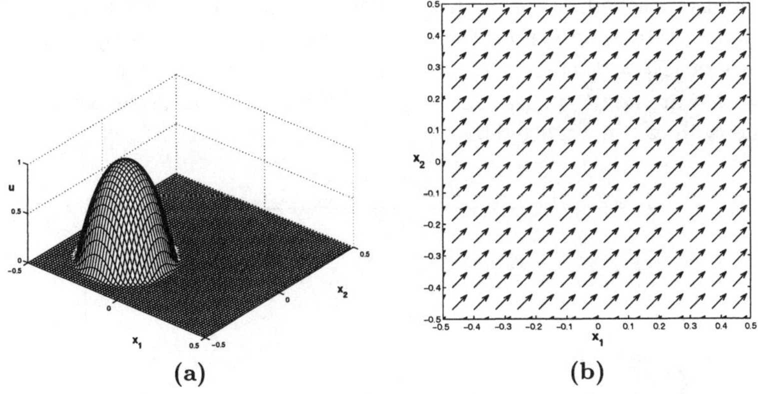

Fig. 6.7. Burgers equation. **(a)** Initial condition u_0 and **(b)** flow field.

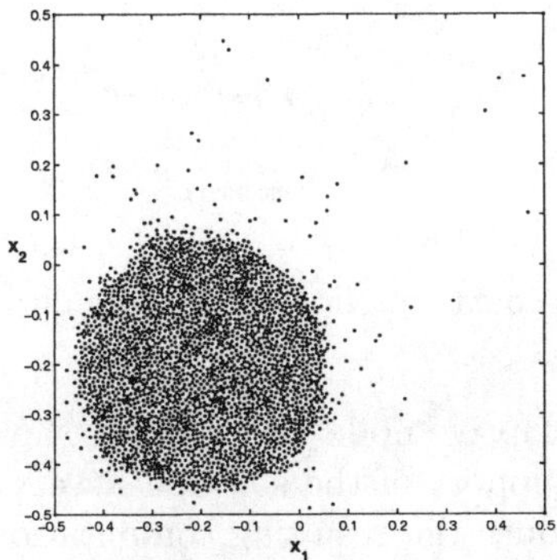

Fig. 6.8. Burgers equation. Initial node set Ξ^0 comprising $|\Xi^0| = 2670$ nodes.

Initially, u_0 is sampled at a set Ξ of $|\Xi| = 1000$ randomly chosen points in the unit square $\Omega = [0,1]^2$. The construction of the initial node set $\Xi^0 \subset \Omega$ is then accomplished by applying the adaption rules of the previous Section 6.3 on the scattered point set Ξ, where we let $\theta_{\text{crs}} = 0.001$ and $\theta_{\text{ref}} = 0.05$ for the relative threshold values, required for the node adaption, see Definition 23 in Section 6.3. The resulting initial node set Ξ^0, of size $|\Xi^0| = 2670$, is displayed in Figure 6.8. Observe that the adaptive distribution of the nodes in Ξ^0 manages to localize the support of the initial condition u_0 very well, see Figure 6.7 (a).

During the simulation, we work with a constant time step size $\tau = 0.004$, and we let $I = [0, 300\tau]$. Moreover, we let $\epsilon = 0.0015$ for the diffusion coefficient, required in Algorithm 22.

A plot of the solution u at three different times, $t_{100} = 100\tau$, $t_{200} = 200\tau$, and $t_{300} = 300\tau$, is shown in Figure 6.10, along with the corresponding node set, of size $|\Xi| = 1256$ at time $t = t_{100}$, $|\Xi| = 1387$ at time $t = t_{200}$, and $|\Xi| = 1383$ at time $t = t_{300}$. The graph of the number of nodes as a function of time is shown in Figure 6.9.

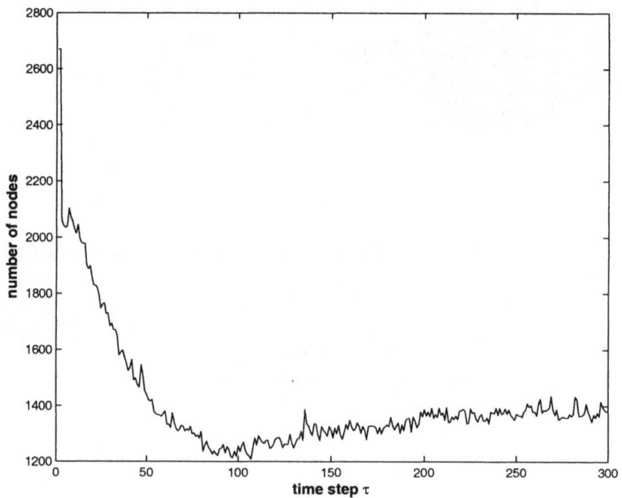

Fig. 6.9. Burgers equation. Number of nodes during the simulation.

Observe that the adaptive node distribution, shown in Figure 6.10, continues to localize the support of the solution u very effectively. This helps, on the one hand, to reduce the resulting computational costs. On the other hand, the shock front propagation is well-resolved by the high node density around the shock, see Figure 6.10. Altogether, the adaptive node distribution manages to capture the evolution of the flow very effectively. This supports the utility of the customized adaption rules.

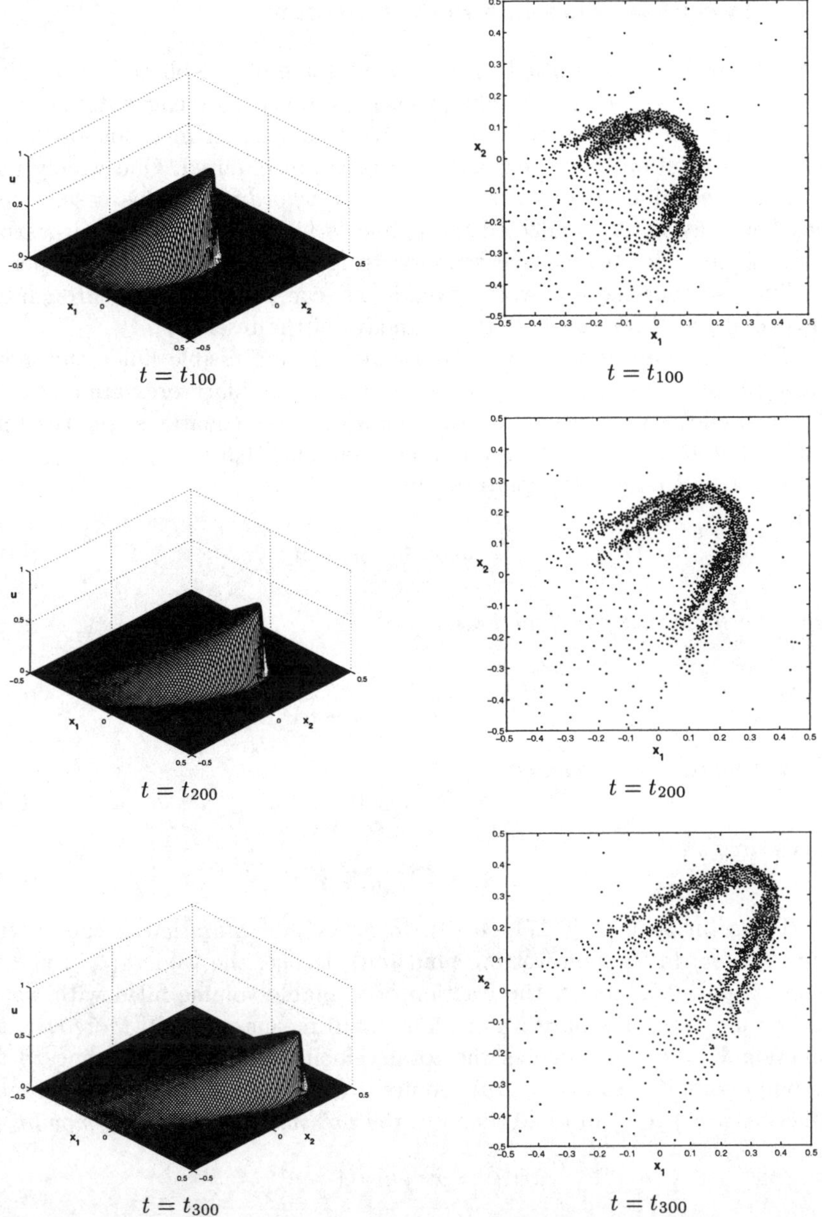

Fig. 6.10. Burgers equation. Evolution of the solution u at three different time steps, $t_{100} = 100\tau$, $t_{200} = 200\tau$, and $t_{300} = 300\tau$ (left column), and the corresponding node distribution (right column).

6.4.3 Two-Phase Flow in a Porous Medium

This subsection is concerned with the simulation of two-phase flow within a porous medium. Two-phase flow problems are typically encountered in applications from *oil reservoir modelling*, where multiscale methods for the numerical simulation of oil recovery processes are required. One widely used standard technique from oil recovery is *waterflooding*. In this case, a wetting fluid, say water, is injected through an *injection well* into a hydrocarbon reservoir in order to displace a non-wetting fluid, say oil, within a porous medium, so that the non-wetting fluid (oil) can be withdrawn through one or several *production wells* at the boundary of the reservoir.

The two-phase flow of two immiscible, incompressible fluids through a homogeneous porous medium in the absence of capillary pressure and gravitational effects is described by the following three equations (see the textbooks [4, 130, 154], and the related discussion in [163]):

The *Buckley-Leverett equation* [20]

$$\frac{\partial u}{\partial t} + \mathbf{v} \cdot \nabla f(u) = 0, \tag{6.17}$$

with the *fractional flow function*

$$f(u) = \frac{u^2}{u^2 + \mu(1-u)^2}, \tag{6.18}$$

the *incompressibility relation*

$$\nabla \cdot \mathbf{v} = 0 \tag{6.19}$$

and *Darcy's law*

$$\mathbf{v} = -M(u)\nabla \mathbf{p}. \tag{6.20}$$

The solution u of (6.17), (6.19), (6.20) is the *saturation* of the wetting fluid (water) in the non-wetting fluid (oil). Hence, the value $u(t,x)$ is, at a point x and at a time t, the fraction of available volume filled with water, and so $u = 1$ means pure water, and $u = 0$ means pure oil. Moreover, the function $\mathbf{v} = \mathbf{v}(t,x)$ denotes the total velocity, which is, according to the incompressibility relation (6.19) required to be divergence-free. Finally, the function $\mathbf{p} = \mathbf{p}(t,x)$ in (6.20) denotes the *pressure*, and the *total mobility*

$$M(u) = u^2 + \mu \cdot (1-u)^2,$$

in (6.20) depends on the *permeability* of the medium and on the ratio μ of the viscosity of the wetting fluid (water) to that of the non-wetting fluid (oil).

We remark that the incompressibility relation (6.19) together with Darcy's law (6.20) form an elliptic equation. The Buckley-Leverett equation (6.17) is a hyperbolic equation, which develops discontinuities in the solution u, even for smooth initial conditions.

The Five-Spot Problem. In order to illustrate the good performance of our adaptive meshfree advection scheme, **AMMoC**, we consider using one popular test case scenario from hydrocarbon reservoir modelling, termed the *five-spot problem*.

The following instance of the five-spot problem may be summarized as follows. The computational domain $\Omega = [-0.5, 0.5]^2$ is corresponding to a bounded reservoir. Initially, the pores of the reservoir are saturated with non-wetting fluid (oil, $u \equiv 0$), before wetting fluid (water, $u \equiv 1$) is injected through one injection well, placed at the center $\mathbf{o} = (0,0)$ of Ω. During the simulation, the non-wetting fluid (oil) is displaced by the wetting fluid (water) towards the four corner points

$$C = \{(-0.5, -0.5), (-0.5, 0.5), (0.5, -0.5), (0.5, 0.5)\}$$

of the domain Ω.

Therefore, the five-spot problem requires solving the equations (6.17),(6.19), (6.20) on Ω, in combination with the initial condition

$$u_0(x) = \begin{cases} 1 & \text{for } \|x - \mathbf{o}\| \leq R, \\ 0 & \text{otherwise,} \end{cases} \tag{6.21}$$

where we let $R = 0.02$ for the radius of the injection well at the center $\mathbf{o} \in \Omega$.

Our aim in this subsection, however, is to merely solve the Buckley-Leverett equation

$$\frac{\partial u}{\partial t} + \mathbf{v} \cdot \nabla f(u) = 0,$$

with f being the fractional flow function in (6.18), with respect to the initial condition (6.21). To this end, we work with the following two simplifications of the five-spot problem.

Firstly, following along the lines of Albright [2], we assume unit mobility, $M \equiv 1$, so that the set (6.19), (6.20) of elliptic equations uncouples from the hyperbolic Buckley-Leverett equation (6.17).

Secondly, we work with a *stationary* pressure field, $\mathbf{p}(x) \equiv \mathbf{p}(\cdot, x)$, given by

$$\mathbf{p}(x) = \sum_{\mathbf{c} \in C} \log(\|x - \mathbf{c}\|) - \log(\|x - \mathbf{o}\|), \qquad \text{for all } x \in \Omega, t \in I, \tag{6.22}$$

which yields the *stationary* velocity field

$$\mathbf{v} = -\nabla \mathbf{p}, \tag{6.23}$$

on the relation (6.20), and with the assumption $M \equiv 1$. It is easy to see that the velocity field $\mathbf{v}$ is in this case divergence-free, i.e., $\mathbf{v}$ in (6.23) satisfies the incompressibility relation (6.19).

Figure 6.11 shows the contour lines of the pressure field $\mathbf{p}$ together with the resulting streamlines of the velocity field $\mathbf{v}$ and the velocity vectors.

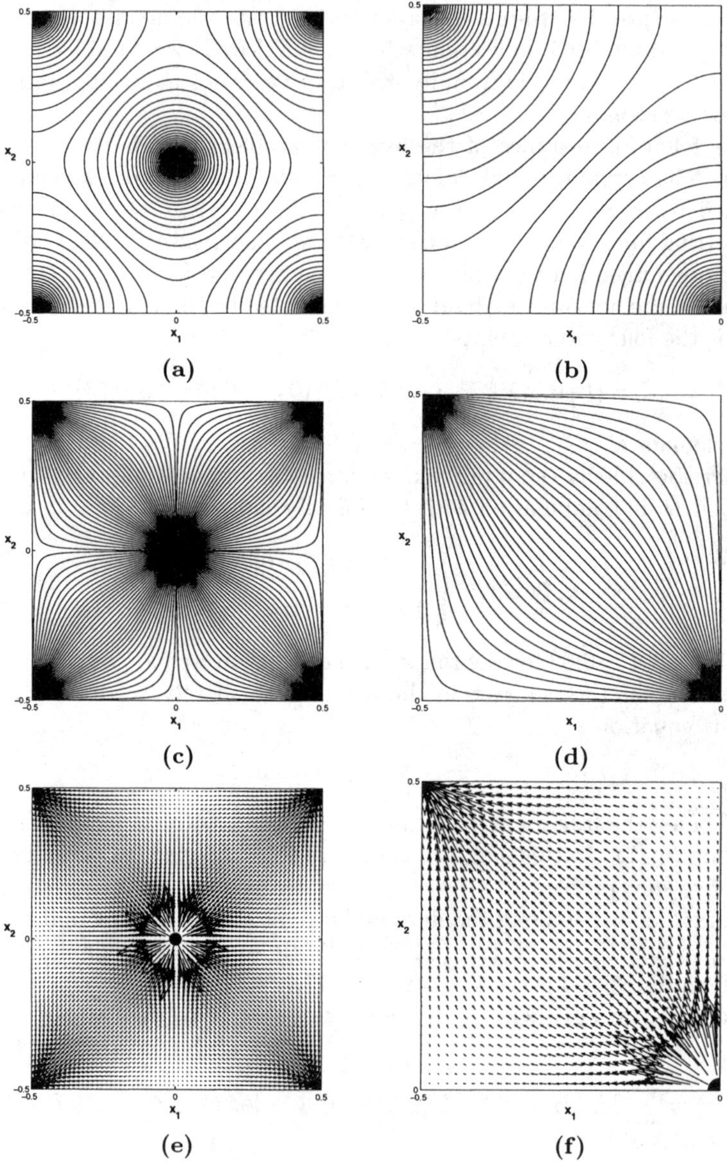

Fig. 6.11. Five-spot problem. **(a)** Contours of the pressure field, **(c)** streamlines of the velocity field, and **(e)** velocity vectors in $\Omega = [-0.5, 0.5]^2$. The corresponding plots of these data in the top left quarter $[-0, 5, 0] \times [0, 0.5]$ are shown in **(b)**, **(d)**, and **(f)**.

As shall be supported by the subsequent numerical comparisons between **AMMoC** and two commercial reservoir simulators, ECLIPSE and FrontSim, the above taken simplifications for the five-spot problem are quite reasonable. In particular, note that the velocity field $\mathbf{p}$ in (6.22) has singularities at the four corners in $\mathcal{C}$ and at the center $\mathbf{o}$, leading to a high velocity $\mathbf{v}$ near the five wells, but small velocity between the wells.

Numerical Results. According to the discussion on **AMMoC** in the previous Section 6.2, we consider numerically solving the viscous Buckley-Leverett equation

$$\frac{\partial u}{\partial t} + \mathbf{v} \cdot \nabla f(u) = \epsilon \cdot \Delta u, \tag{6.24}$$

rather than the hyperbolic (6.17), in combination with the initial condition (6.21).

Recall that this modification of (6.17) is required in order to model the propagation of the shock front, which is occurring at the interface between the two fluids, water and oil. We remark that the nature and location of the shock front between the two interfaces is of primary importance in the relevant application, oil reservoir simulation. So is for instance the *breakthrough time*, the time required for the wetting fluid to arrive at the production well, of particular interest. Therefore, the accurate approximation of the shock front requires particular care. We accomplish this by the adaptive distribution of the nodes during the simulation.

Now let us turn to our numerical results, provided by **AMMoC**. In our simulation, we decided to select a constant time step size $\tau = 5 \cdot 10^{-5}$, and the simulation comprises 2100 time steps, so that $I = [0, 2100\tau]$. Moreover, we let $\epsilon = 0.015$ for the diffusion coefficient in (6.24), and $\theta_{\text{crs}} = 0.005$ and $\theta_{\text{ref}} = 0.05$ for the relative threshold values, required for the node adaption, see Definition 23 in Section 6.3. Finally, we selected the value $\mu = 0.5$ for the viscosity ratio of water and oil, appearing in the fractional flow function (6.18).

Figure 6.12 displays the water saturation u during the simulation at six different times, $t = t_0, t = t_{120}, t = t_{240}, t = t_{360}, t = t_{480}$, and $t = t_{600}$, where u is evaluated at a fixed cartesian mesh comprising 100×100 rectangular cells. The corresponding color code for the water saturation is shown at the right margin of Figure 6.12, respectively.

Note that the shock front, at the interface between the non-wetting fluid (oil, $u \equiv 0$) and the wetting fluid (water, $u \equiv 1$), is moving from the center towards the four corner points of the computational domain Ω. This way, the non-wetting fluid (oil) is effectively displaced by the wetting fluid (water) into the four production wells, as expected.

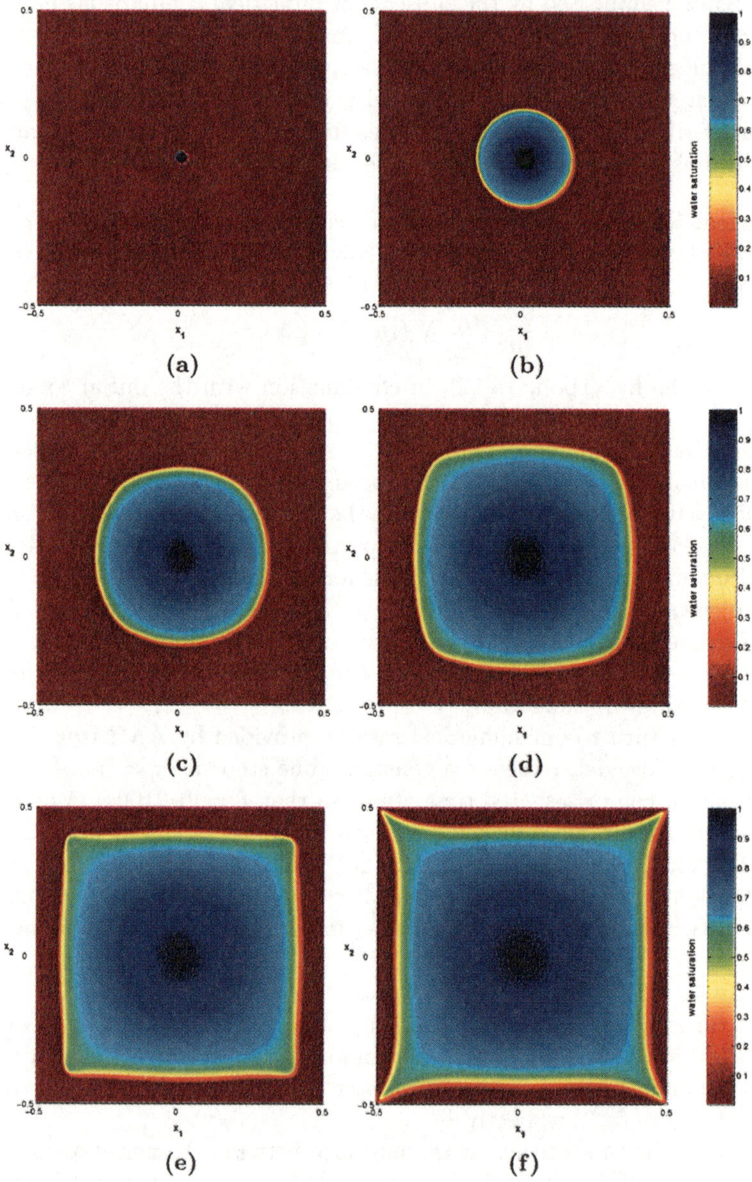

Fig. 6.12. Five-spot problem. Solution obtained by **AMMoC**. The color plots indicate the water saturation u during the simulation at six different times, **(a)** $t = t_0$, **(b)** $t = t_{120}$, **(c)** $t = t_{240}$, **(d)** $t = t_{360}$, **(e)** $t = t_{480}$, and **(f)** $t = t_{600}$.

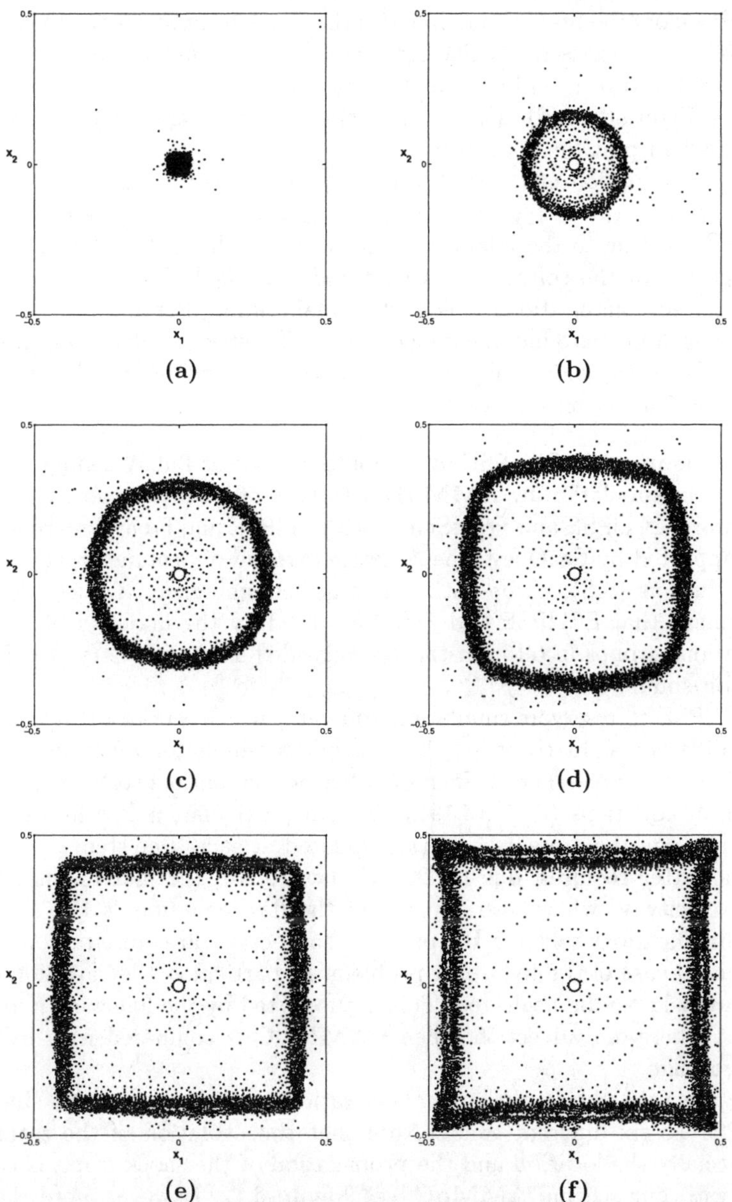

Fig. 6.13. Five-spot problem. Adaptive node distribution during the simulation by **AMMoC** at six different times, **(a)** $t = t_0$, **(b)** $t = t_{420}$, **(c)** $t = t_{840}$, **(d)** $t = t_{1260}$, **(e)** $t = t_{1680}$, and **(f)** $t = t_{2100}$.

Just before the breakthrough, when the shock front arrives at the production wells, an increased velocity can be observed around the four production wells, see the transition between the time step $t = t_{480}$, Figure 6.12 **(e)** and $t = t_{600}$, Figure 6.12 **(f)**. This *sucking effect* is due to the singularities of the pressure field $\mathbf{p}$ at the corners in $\mathcal{C}$.

The distribution of the nodes, corresponding to the six different times, $t = t_0$, $t = t_{120}$, $t = t_{240}$, $t = t_{360}$, $t = t_{480}$, and $t = t_{600}$, is displayed in Figure 6.13. Due to the adaptive distribution of the nodes, the shock front propagation of the solution u is captured very well. This helps to reduce the required computational costs while maintaining the accuracy, due to a higher resolution around the shock front. The effective distribution of the nodes around the shock supports the utility of the adaption rules, proposed in Section 6.3, yet once more.

Comparison with ECLIPSE and FrontSim. Let us finally compare the results of our simulation by **AMMoC** with two different commercial reservoir simulators, ECLIPSE and FrontSim. Both ECLIPSE and FrontSim are licensed and supported by GeoQuest, the software division of Schlumberger Information Solutions (SIS), an operating unit of Schlumberger Oil Field Services. We remark that ECLIPSE and FrontSim, used by the majority of reservoir simulations groups in oil industry, are regarded as the industry standard in reservoir simulation.

ECLIPSE is reservoir simulation software, which works with *first order* finite differences. In contrast, the multiphase simulator FrontSim is based on a *front tracking* scheme. Each of these two simulators solves the coupled system of equations (6.17),(6.19),(6.20). In particular, unlike in our model simplification, the pressure field $\mathbf{p}$ is updated at each time step.

The latter requires, due to Darcy's law (6.20), the maintenance of the total velocity $\mathbf{v}$, which also appears in the flow equation (6.17). However, our simplifications taken in the previous subsection, are quite reasonable for the special case of the five-sport problem. In particular, the variation of the pressure field can be neglected. This is supported by the following numerical results, where our advection scheme **AMMoC** is compared with ECLIPSE and FrontSim.

Figure 6.15 shows the water saturation obtained from the simulator ECLIPSE at six different times. Note that the evolution of the saturation u, especially the location and the propagation of the shock front, is comparable with our scheme **AMMoC**, see Figure 6.12. However, note that the resolution of the shock front (at the interface of the two fluids) obtained by our scheme **AMMoC** in Figure 6.12 is much *sharper* than the shock front obtained by ECLIPSE in Figure 6.15.

As regards the corresponding results obtained by the simulator FrontSim, these are displayed in Figure 6.16. Note that FrontSim widely avoids numerical diffusion around the shock front, and so the interface between the two fluids is resolved very sharply.

For the purpose of further comparison, we considered recording the water saturation u at time $t = t_{1260}$, for each of the three different simulation methods, ECLIPSE, FrontSim, and **AMMoC**. Figure 6.14 shows the three different profiles of the saturation $u(t_{1260}, \cdot)$ across the half diagonal of Ω, drawn from the center $\mathbf{o} = (0,0)$ to the corner point $(0.5, 0.5)$. For better orientation, the dashed line in Figure 6.14 shows the expected height of the shock front, which can be computed analytically by Welge's tangent method [169].

Note that the three different methods lead to similar saturation profiles. Moreover, each method captures the expected height of the shock front fairly well. When it comes to accurately resolving the shock front, the method FrontSim is the best, followed by our meshfree scheme **AMMoC** and lastly ECLIPSE. This is not very surprising insofar as FrontSim relies on *front tracking*, a technique which is well-known for its small numerical diffusion.

Since the method ECLIPSE is only of first order, ECLIPSE is inferior to both **AMMoC** and FrontSim, due to enhanced numerical diffusion around the shock front. Our meshfree advection scheme **AMMoC**, of second order, reduces (compared with ECLIPSE) the numerical diffusion, mainly due to the effective adaptive node distribution. Moreover, the saturation profile obtained by our meshfree method **AMMoC** is fairly close that of FrontSim, see Figure 6.14.

Altogether, we feel that our meshfree method **AMMoC** is, as regards its performance concerning the five-spot problem, quite competitive.

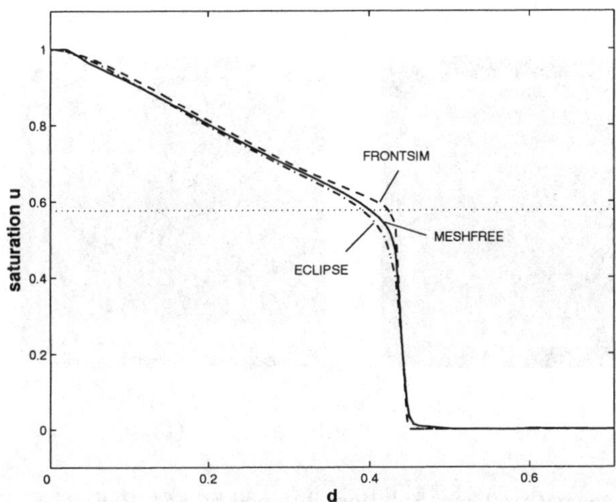

Fig. 6.14. Five-spot problem. Comparison between ECLIPSE, FrontSim, and our meshfree advection scheme **AMMoC**. The saturation profiles are, at time $t = t_{1260}$, compared along the half diagonal from the center $\mathbf{o} = (0,0)$ to the corner point $(0.5, 0.5)$. The values on the axis of abscissae correspond to the distance d from $\mathbf{o}$.

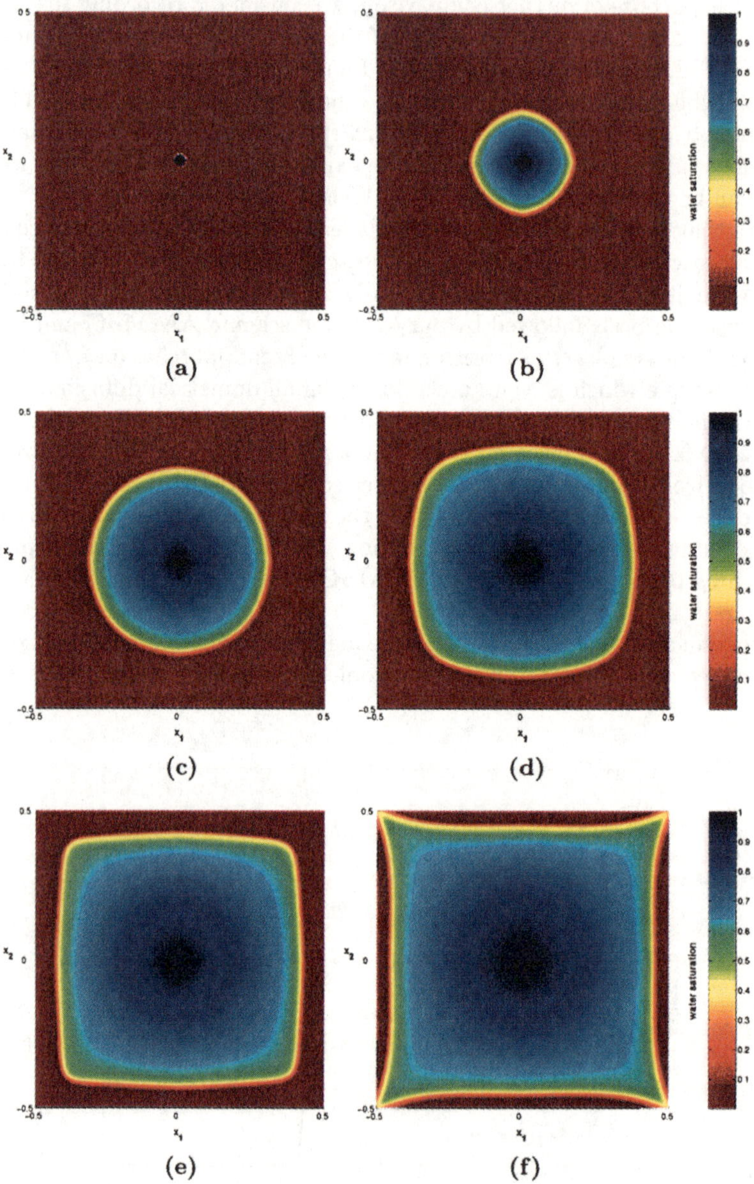

Fig. 6.15. Five-spot problem. Solution obtained by ECLIPSE. The color plots indicate the water saturation u during the simulation at six different times, **(a)** $t = t_0$, **(b)** $t = t_{420}$, **(c)** $t = t_{840}$, **(d)** $t = t_{1260}$, **(e)** $t = t_{1680}$, and **(f)** $t = t_{2100}$.

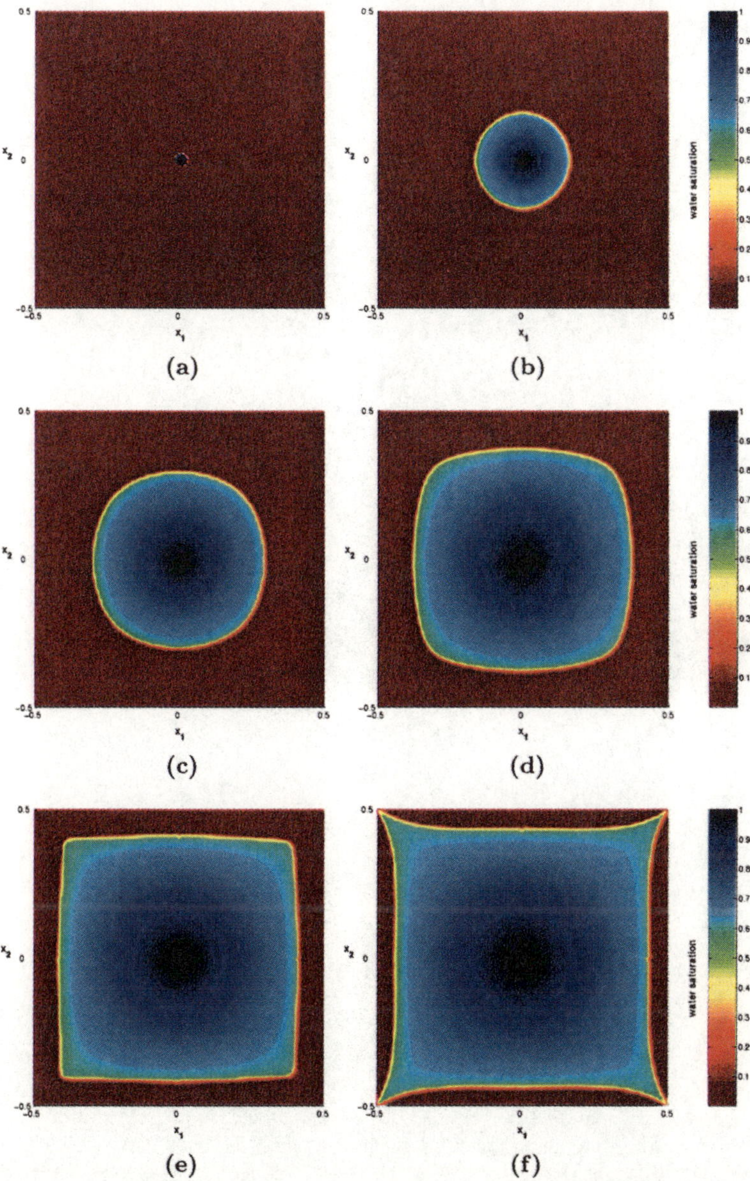

Fig. 6.16. Five-spot problem. Solution obtained by FrontSim. The color plots indicate the water saturation u during the simulation at six different times, **(a)** $t = t_0$, **(b)** $t = t_{420}$, **(c)** $t = t_{840}$, **(d)** $t = t_{1260}$, **(e)** $t = t_{1680}$, and **(f)** $t = t_{2100}$.

Bibliography

1. M. Abramowitz and I.A. Stegun (1970) *Handbook of Mathematical Functions*. Dover Publications, New York.
2. N. Albright, P. Concus, and W. Proskurowski (1979) Numerical solution of the multi-dimensional Buckley-Leverett equation by a sampling method. Paper SPE 7681, Soc. Petrol. Eng. Fifth Symp. on Reservoir Simulation, Denver, Feb. 1979.
3. A. Aldroubi and K. Gröchenig (2001) Nonuniform sampling and reconstruction in shift-invariant spaces. SIAM Review **43**, 585–620.
4. K. Aziz and A. Settari (1979) *Petroleum Reservoir Simulation*. Applied Science, London.
5. I. Babuška and J.M. Melenk (1997) The partition of unity method. Int. J. Numer. Meths. Eng. **40**, 727–758.
6. K. Ball (1992) Eigenvalues of Euclidean distance matrices. J. Approx. Theory **68**, 74–82.
7. K. Ball, N. Sivakumar, and J.D. Ward (1992) On the sensitivity of radial basis interpolation to minimal distance separation. J. Approx. Theory **8**, 401–426.
8. J. Behrens (1996) An adaptive semi-Lagrangian advection scheme and its parallelization. Mon. Wea. Rev. **124**, 2386–2395.
9. J. Behrens, K. Dethloff, W. Hiller, and A. Rinke (2000) Evolution of small scale filaments in an adaptive advection model for idealized tracer transport. Mon. Wea. Rev. **128**, 2976–2982.
10. J. Behrens and A. Iske (2002) Grid-free adaptive semi-Lagrangian advection using radial basis functions. Comput. Math. Appl. **43**, 319–327.
11. J. Behrens, A. Iske, and S. Pöhn (2001) Effective node adaption for grid-free semi-Lagrangian advection. *Discrete Modelling and Discrete Algorithms in Continuum Mechanics*, T. Sonar and I. Thomas (eds.), Logos Verlag, Berlin, 110–119.
12. J. Behrens, A. Iske, and M. Käser (2002) Adaptive meshfree method of backward characteristics for nonlinear transport equations. *Meshfree Methods for Partial Differential Equations*, M. Griebel and M.A. Schweitzer (eds.), Springer-Verlag, Heidelberg, 21–36.
13. S. Bernstein (1928) Sur les fonctions absolument monotones. Acta Mathematica **51**, 1-66.
14. Å. Bjørck: *Numerical Methods for Least Squares Problems*, SIAM, Philadelphia, 1996.
15. Å. Bjørck and G.H. Golub (1967) Iterative refinement of linear least squares solutions by Householder transformation. BIT **7**, 322–337.

16. G.-P. Bonneau (2002) BLaC wavelets & non-nested wavelets. *Tutorials on Multiresolution in Geometric Modelling*, A. Iske, E. Quak, and M.S. Floater (eds.), Springer-Verlag, Heidelberg, 147–163.
17. C. de Boor, K. Höllig, and S.D. Riemenschneider (1993) *Box-Splines.* Springer-Verlag, New York.
18. J. Bramble and M. Zlamal (1970) Triangular elements in the finite element method. Math. Comp. **24**, 809–820.
19. A. Bregman, A.F.J. van Velthoven, F.G. Wienhold, H. Fischer, T. Zenker, A. Waibel, A. Frenzel, F. Arnold, G.W. Harris, M.J.A. Bolder, and J. Lelieveld (1995) Aircraft measurements of O_3, HNO_3, and N_2O in the winter arctic lower stratosphere during the stratosphere-troposphere experiment by aircraft measurements (STREAM) 1. J. Geophys. Res. **100**, 11245–11260.
20. J.M. Buckley and M.C. Leverett (1942) Mechanism of fluid displacement in sands. Trans. AIME **146**, 107–116.
21. M.D. Buhmann (1998) Radial functions on compact support. Proc. Edinburgh Math. Soc. **41**, 33–46.
22. M.D. Buhmann (2000) Radial basis functions. Acta Numerica, 1–38.
23. M.D. Buhmann (2003) *Radial basis functions: theory and implementations.* Cambridge University Press, Cambridge, UK.
24. Y.D. Burago and V. A. Zalgaller (1988) *Geometric Inequalities.* Springer, Berlin.
25. J.M. Burgers (1940) Application of a model system to illustrate some points of the statistical theory of free turbulence. Proc. Acad. Sci. Amsterdam **43**, 2–12.
26. J.M. Carnicer and M.S. Floater (1996) Piecewise linear interpolants to Lagrange and Hermite convex scattered data. Numer. Algorithms **13**, 345–364.
27. E. Catmull and J. Clark (1978) Recursively generated B-spline surfaces on arbitrary topological meshes. Computer-Aided Design **10**, 350–355 (reprinted in Seminal Graphics, Wolfe (ed.), SIGGRAPH 1998).
28. A. Cohen and R. Ryan (1995) *Wavelets and multiscale signal processing.* Chapman and Hall, London.
29. A. Cohen, W. Dahmen, I. Daubechies, and R. DeVore (2001) Tree approximation and optimal encoding, Appl. Comput. Harmonic Anal. **11**, 192–226.
30. A. Cohen and B. Matei (2002) Nonlinear subdivision schemes: applications to image processing. *Tutorials on Multiresolution in Geometric Modelling*, A. Iske, E. Quak, and M.S. Floater (eds.), Springer-Verlag, Heidelberg, 93–97.
31. M. Conrad and J. Prestin (2002) Multiresolution on the sphere. *Tutorials on Multiresolution in Geometric Modelling*, A. Iske, E. Quak, and M.S. Floater (eds.), Springer-Verlag, Heidelberg, 165–202.
32. J.H. Conway and N.J.A. Sloane (1993) *Sphere Packings, Lattices and Groups.* Springer, New York.
33. T.H. Cormen, C.E. Leiserson, R.L. Rivest, and C. Stein (2001) *Introduction to Algorithms*, 2nd edition. MIT Press, Cambridge, Massachusetts.

34. R. Courant, E. Isaacson, and M. Rees (1952). On the solution of nonlinear hyperbolic differential equations by finite differences. Comm. Pure Appl. Math. **5**, 243–255.
35. W. Dahmen and C.A. Micchelli (1988) Convexity of multivariate Bernstein polynomials and box spline surfaces. Stud. Sci. Math. Hung. **23**, 265–287.
36. I. Daubechies (1992) *Ten lectures on wavelets.* SIAM, Philadelphia.
37. G.M. Davis and A. Nosratinia (1999) Wavelet-based image coding: an overview. *Applied and Computational Control, Signals, and Circuits*, B.N. Datta (ed.), Birkhauser, Boston, 205–269.
38. L. De Floriani and P. Magillo (2002) Multiresolution mesh representation: models and data structures. *Tutorials on Multiresolution in Geometric Modelling*, A. Iske, E. Quak, and M.S. Floater (eds.), Springer-Verlag, Heidelberg, 363–417.
39. B. Delaunay (1934) Sur la sphère vide. Bull. Acad. Sci. USSR (VII), Classe Sci. Mat. Nat., 793–800.
40. L. Demaret, N. Dyn, M.S. Floater, and A. Iske (2003) Adaptive thinning for terrain modelling and image compression. Manuscript, Technische Universität München, November 2003.
41. L. Demaret and A. Iske (2003) Scattered data coding in digital image compression. *Curves and Surface Fitting: Saint-Malo 2002.* A. Cohen, J.-L. Merrien, and L.L. Schumaker (eds.), Nashboro Press, Brentwood, 107–117.
42. L. Demaret and A. Iske (2003) Advances in digital image compression by adaptive thinning. To appear in the *MCFA Annals*, Volume III.
43. K. Dethloff, A. Rinke, R. Lehmann, J.H. Christensen, M. Botzet, and B. Machenhauer (1996) Regional climate model of the arctic atmosphere. J. Geophys. Res. **101**, 23401–23422.
44. P. Deuflhard and F. Bornemann (2002) *Scientific Computing with Ordinary Differential Equations.* Springer, New York.
45. D.W.H. Doo and M.A. Sabin (1978) Behaviour of recursive subdivision surfaces near extraordinary points. Computer-Aided Design **10**, 356–360 (reprinted in Seminal Graphics, Wolfe (ed.), SIGGRAPH 1998).
46. J. Duchon (1976) Interpolation des Fonctions de deux variables suivant le principe de la flexion des plaques minces, R.A.I.R.O. Analyse Numeriques **10**, 5–12.
47. J. Duchon (1977) Splines minimizing rotation-invariant semi-norms in Sobolev spaces. *Constructive Theory of Functions of Several Variables*, W. Schempp and K. Zeller (eds.), Springer, Berlin, 85–100.
48. J. Duchon (1978) Sur l'erreur d'interpolation des fonctions de plusieurs variables par les D^m-splines. R.A.I.R.O. Analyse Numeriques **12**, 325–334.
49. D.R. Durran (1999) *Numerical Methods for Wave Equations in Geophysical Fluid Dynamics.* Springer, New York.
50. N. Dyn (1987) Interpolation of scattered data by radial functions. *Topics in Multivariate Approximation*, C.K. Chui, L.L. Schumaker, F.I. Utreras (eds.), Academic Press, 47–61.
51. N. Dyn (1989) Interpolation and approximation by radial and related functions. *Approximation Theory VI*, C. K. Chui, L. L. Schumaker, and J. D. Ward (eds.), Academic Press, New York, 1989, 211–234.

52. N. Dyn (2002) Interpolatory subdivision schemes. *Tutorials on Multiresolution in Geometric Modelling*, A. Iske, E. Quak, and M.S. Floater (eds.), Springer-Verlag, Heidelberg, 25–50.
53. N. Dyn (2002) Analysis of convergence and smoothness by the formalism of Laurent polynomials. *Tutorials on Multiresolution in Geometric Modelling*, A. Iske, E. Quak, and M.S. Floater (eds.), Springer-Verlag, Heidelberg, 51–68.
54. N. Dyn, M.S. Floater, and A. Iske (2001) Univariate adaptive thinning. *Mathematical Methods for Curves and Surfaces: Oslo 2000*, T. Lyche and L.L. Schumaker (eds.), Vanderbilt University Press, Nashville, 123–134.
55. N. Dyn, M.S. Floater, and A. Iske (2002) Adaptive thinning for bivariate scattered data. J. Comput. Appl. Math. **145**(2), 505–517.
56. N. Dyn, J.A. Gregory, and D. Levin (1990) A butterfly subdivision scheme for surface interpolation with tension control. ACM Trans. on Graphics **9**, 160–169.
57. N. Dyn and D. Levin (2002) Subdivision schemes in geometric modelling, Acta Numerica, 73-144.
58. N. Dyn, D. Levin, and S. Rippa (1990) Algorithms for the construction of data dependent triangulations. *Algorithms for Approximation II*, M.G. Cox and J.C. Mason (eds.), Clarendon Press, Oxford, 185–192.
59. N. Dyn, D. Levin, and S. Rippa (1990) Data dependent triangulations for piecewise linear interpolation. IMA J. Numer. Anal. **10**, 137–154.
60. N. Dyn, D. Levin, and S. Rippa (1992) Boundary correction for piecewise linear interpolation defined over data-dependent triangulations. J. Comput. Appl. Math. **39**, 179–192.
61. N. Dyn and S. Rippa (1993) Data-dependent triangulations for scattered data interpolation and finite element approximation. Appl. Numer. Math. **12**, 89–105.
62. M. Falcone and R. Ferretti (1998) Convergence analysis for a class of high-order semi-Lagrangian advection schemes. SIAM J. Numer. Anal. **35**, 909–940.
63. G.E. Fasshauer (1999) On smoothing for multilevel approximation with radial basis functions. *Approximation Theory XI, Vol.II: Computational Aspects*, C.K. Chui and L.L. Schumaker (eds.), Vanderbilt University Press, 55–62.
64. G.E. Fasshauer (1999) Solving differential equations with radial basis functions: multilevel methods and smoothing, Advances in Comp. Math. **11**, 139–159.
65. G.E. Fasshauer and J.W. Jerome (1999) Multistep approximation algorithms: improved convergence rates through postconditioning with smoothing kernels. Advances in Comp. Math. **10**, 1–27.
66. G.E. Fasshauer, E.C. Gartland jr., and J.W. Jerome (2000) Algorithms defined by Nash iteration: some implementations via multilevel collocation and smoothing J. Comput. Appl. Math. **119**, 161–183.
67. T. Feder and D.H. Greene (1988) Optimal algorithms for approximate clustering. *Proceedings of the 20th Annual ACM Symposium on Theory of Computing*, Chicago, Illinois, 434–444.

68. H.G. Feichtinger and K. Gröchenig (1989) Multidimensional irregular sampling of band-limited functions in L^p-spaces. International Series of Numerical Mathematics **90**, 135–142.
69. H.G. Feichtinger and K. Gröchenig (1992) Iterative reconstruction of multivariate band-limited functions from irregular sampling values. SIAM J. Math. Anal. **23**, 244–261.
70. H.G. Feichtinger and K. Gröchenig (1993) Error analysis in regular and irregular sampling theory. Applicable Analysis **50**, 167–189.
71. R.A. Finkel and J.L. Bentley (1974) Quad-trees; a data structure for retrieval on composite keys. Acta Inform. **4**, 1–9.
72. M.S. Floater and A. Iske (1996) Multistep scattered data interpolation using compactly supported radial basis functions. J. Comput. Appl. Math. **73**, 65–78.
73. M.S. Floater and A. Iske (1996) Thinning and approximation of large sets of scattered data. *Advanced Topics in Multivariate Approximation*, F. Fontanella, K. Jetter and P.-J. Laurent (eds.), World Scientific, Singapore, 87–96.
74. M.S. Floater and A. Iske (1998) Thinning algorithms for scattered data interpolation. BIT **38**, 705–720.
75. M.S. Floater and E. Quak (1999) Piecewise linear prewavelets on arbitrary triangulations. Numer. Math. **82**, 221–252.
76. M.S. Floater, E. Quak, and M. Reimers (2000) Filter bank algorithms for piecewise linear prewavelets on arbitrary triangulations. J. Comput. Appl. Math. **119**, 185–207.
77. J. Fürst and T. Sonar (2001) On meshless collocation approximations of conservation laws: positive schemes and dissipation models. Z. Angew. Math. Mech. **81**, 403–415.
78. I.M. Gelfand and G.E. Shilov (1964) *Generalized Functions, Vol. 1: Properties and Operations*. Academic Press, New York.
79. I.M. Gelfand and N.Y. Vilenkin (1964) *Generalized Functions, Vol. 4: Applications of Harmonic Analysis*. Academic Press, New York.
80. C. Gotsman, S. Gumhold, and L. Kobbelt (2002) Simplification and Compression of 3D Meshes. *Tutorials on Multiresolution in Geometric Modelling*, A. Iske, E. Quak, and M.S. Floater (eds.), Springer-Verlag, Heidelberg, 319–361.
81. J.A. Gregory (1975) Error bounds for linear interpolation in triangles. *The Mathematics of finite elements and applications II*, J. R. Whiteman (ed.), Academic Press, London, 163–170.
82. K. Guo, S. Hu, and X. Sun (1993) Conditionally positive definite functions and Laplace-Stieltjes integrals. J. Approx. Theory **74**, 249–265.
83. B. Gustafsson, H.-O. Kreiss, and J. Oliger (1995) *Time Dependent Problems and Difference Methods*. John Wiley and Sons, New York.
84. T. Gutzmer and A. Iske (1997) Detection of discontinuities in scattered data approximation. Numer. Algorithms **16**, 155–170.
85. R.L. Hardy (1971) Multiquadric equations of topography and other irregular surfaces. J. Geophys. Res. **76**, 1905–1915.
86. P.S. Heckbert and M. Garland (1997) Survey of surface simplification algorithms. Technical Report, Computer Science Dept., Carnegie Mellon University.

87. N.J. Higham (1996) *Accuracy and Stability of Numerical Algorithms.* SIAM, Philadelphia.
88. D.S. Hochbaum (1997) *Approximation algorithms for NP-hard problems.* PWS Publishing Company, Boston.
89. D.S. Hochbaum and D.B. Shmoys (1985) A best possible heuristic for the k-center problem. *Mathematics of Operations Research* **10**, 180–184.
90. D.A. Huffman (1952) A method for the construction of minimum-redundancy codes. Proc. Inst. Radio Eng., 1098–1101.
91. A. Iske (1994) *Charakterisierung bedingt positiv definiter Funktionen für multivariate Interpolationsmethoden mit radialen Basisfunktionen.* Dissertation, Universität Göttingen.
92. A. Iske (1995) Characterization of function spaces associated with conditionally positive definite functions. *Mathematical Methods for Curves and Surfaces*, M. Dæhlen, T. Lyche, and L. L. Schumaker (eds.), Vanderbilt University Press, Nashville, 1995, 265–270.
93. A. Iske (1995) Reconstruction of functions from generalized Hemite-Birkhoff data. *Approximation Theory VIII, Vol. 1: Approximation and Interpolation*, C. K. Chui and L. L. Schumaker (eds.), World Scientific, Singapore, 1995, 257–264.
94. A. Iske (1999) Reconstruction of smooth signals from irregular samples by using radial basis function approximation. *Proceedings of the 1999 International Workshop on Sampling Theory and Applications*, Y. Lyubarskii (ed.), The Norwegian University of Science and Technology, Trondheim, 82–87.
95. A. Iske (2000) Optimal distribution of centers for radial basis function methods. Report, Technische Universität München.
96. A. Iske (2001) Hierarchical scattered data filtering for multilevel interpolation schemes. *Mathematical Methods for Curves and Surfaces: Oslo 2000*, T. Lyche and L.L. Schumaker (eds.), Vanderbilt University Press, Nashville, 211–220.
97. A. Iske (2002) Scattered data modelling using radial basis functions. *Tutorials on Multiresolution in Geometric Modelling*, A. Iske, E. Quak, and M.S. Floater (eds.), Springer-Verlag, Heidelberg, 205–242.
98. A. Iske and J. Levesley (2002) Multilevel scattered data approximation by adaptive domain decomposition. Technical Report, University of Leicester.
99. A. Iske, E. Quak, and M.S. Floater (2002) *Tutorials on Multiresolution in Geometric Modelling.* Springer-Verlag, Heidelberg.
100. A. Iske (2003) Adaptive irregular sampling in meshfree flow simulation. *Sampling, Wavelets, and Tomography*, J. Benedetto and A. Zayed (eds.), Applied and Numerical Harmonic Analysis Series, Birkhäuser, Boston.
101. A. Iske (2003) Progressive scattered data filtering. J. Comput. Appl. Math. **158**(2), 297–316.
102. A. Iske (2003) On the approximation order and numerical stability of local Lagrange interpolation by polyharmonic splines. *Modern Developments in Multivariate Approximation*, W. Haussmann, K. Jetter, M. Reimer, J. Stöckler (eds.), ISNM 145, Birkhäuser, Basel, 153–165.

103. A. Iske (2003) Radial basis functions: basics, advanced topics, and meshfree methods for transport problems. To appear in *Rendiconti del Seminario Matematico*, Università e Politecnico di Torino.
104. A. Iske and M. Käser (2003) Conservative semi-Lagrangian advection on adaptive unstructured meshes. To appear in *Numerical Methods for Partial Differential Equations*.
105. H. Jung (1901) Über die kleinste Kugel, die eine räumliche Figur einschließt. J. Reine Angew. Math. **123**, 241–257.
106. M. Käser (2003) *Adaptive Methods for the Numerical Simulation of Transport Processes*. Dissertation, Technische Universität München.
107. O. Kariv and S.L. Hakimi (1979) An algorithmic approach to network location problems, part I: the p-centers. SIAM J. Appl. Math. **37**, 513–538.
108. J.P.R. Laprise and A. Plante (1995) A class of semi-Lagrangian integrated-mass (SLIM) numerical transport algorithms. Monthly Weather Review **123**, 553–565.
109. C.L. Lawson (1972) Generation of a triangular grid with application to contour plotting, JPL 299.
110. C.L. Lawson (1977) Software for C^1 interpolation. *Mathematical Software III*, J. Rice (ed.), Academic Press, New York, 161–194.
111. C.L. Lawson and R.J. Hanson (1974) *Solving Least Squares Problems*. Prentice-Hall, Englewood Cliffs, N.J.
112. E. Le Pennec and S. Mallat (2000) Image compression with geometrical wavelets, Proceedings of ICIP 2000, September 2000, Vancouver.
113. R.L. LeVeque (2002) *Finite Volume Methods for Hyperbolic Problems*. Cambridge University Press, Cambridge, UK.
114. W.R. Madych and S.A. Nelson (1983) Multivariate interpolation: a variational theory. Manuscript.
115. W.R. Madych and S.A. Nelson (1988) Multivariate interpolation and conditionally positive definite functions. Approx. Theory Appl. **4**, 77–89.
116. W.R. Madych and S.A. Nelson (1990) Multivariate interpolation and conditionally positive definite functions II, Math. Comp. **54**, 211–230.
117. J. Meinguet (1979) Multivariate interpolation at arbitrary points made simple. Z. Angew. Math. Phys. **30**, 292–304.
118. J. Meinguet (1979) An intrinsic approach to multivariate spline interpolation at arbitrary points. *Polynomial and Spline Approximations*, N.B. Sahney (ed.), Reidel, Dordrecht, 163–190.
119. J. Meinguet (1984) Surface spline interpolation: basic theory and computational aspects. *Approximation Theory and Spline Functions*, S.P. Singh, J.H. Bury, and B. Watson (eds.), Reidel, Dordrecht, 127–142.
120. J.M. Melenk and I. Babuška (1996) The partition of unity finite element method: basic theory and applications. Comput. Meths. Appl. Mech. Engrg. **139**, 289–314.
121. C.A. Micchelli (1986) Interpolation of scattered data: distance matrices and conditionally positive definite functions. Constr. Approx. **2**, 11–22.
122. C.A. Micchelli, T.J. Rivlin, and S. Winograd (1976) Optimal recovery of smooth functions. Numer. Math. **260**, 191–200.

123. J.J Monaghan (1992) Smoothed particle hydrodynamics. Ann. Rev. Astron. and Astrophysics **30**, 543–574.
124. K.W. Morton (1996) *Numerical Solution of Convection-Diffusion Problems.* Chapman & Hall, London.
125. B. Mulansky (1990) Interpolation of scattered data by a bivariate convex function, I: piecewise linear C^0 interpolation. Memorandum **858**, Faculty of Applied Mathematics, University of Twente, Enschede.
126. F.J. Narcowich, R. Schaback, and J.D. Ward (1999) Multilevel interpolation and approximation. Appl. Comput. Harmonic Anal. **7**, 243–261.
127. F.J. Narcowich and J.D. Ward (1991) Norms of inverses and conditions numbers for matrices associated with scattered data. J. Approx. Theory **64**, 69–94.
128. F.J. Narcowich and J.D. Ward (1992) Norm estimates for the inverses of a general class of scattered-data radial-function interpolation matrices. J. Approx. Theory **69**, 84–109.
129. J. O'Rourke, C.-B. Chien, T. Olson, and D. Naddor (1982) A new linear algorithm for intersecting convex polygons. Comput. Graph. Image Process. **19**, 384–391.
130. D.W. Peaceman (1977) *Fundamentals of Numerical Reservoir Simulation.* Elsevier, Amsterdam.
131. T.N. Phillips and A. J. Williams (2001) Conservative semi-Lagrangian finite volume schemes. Numer. Meth. Part. Diff. Eq. **17**, 403–425.
132. S. Pöhn (2001) *Implementierung eines gitterfreien adaptiven Advektionsschemas.* Diplomarbeit, Fakultät für Informatik, Technische Universität München.
133. M.J.D. Powell (1992) The theory of radial basis function approximation in 1990. *Advances in Numerical Analysis II: Wavelets, Subdivision, and Radial Basis Functions*, W.A. Light (ed.), Clarendon Press, Oxford, 105–210.
134. F.P. Preparata and M.I. Shamos (1988) *Computational Geometry.* Springer, New York.
135. A. Priestley (1993) A quasi-conservative version of the semi-Lagrangian advection scheme. Monthly Weather Review **121**, 621–629.
136. E. Quak (2002) Nonuniform B-splines and B-wavelets. *Tutorials on Multiresolution in Geometric Modelling*, A. Iske, E. Quak, and M.S. Floater (eds.), Springer-Verlag, Heidelberg, 101–146.
137. E. Quak and L.L. Schumaker (1990) Cubic spline fitting using data dependent triangulations. Comput. Aided Geom. Design **7**, 293–301.
138. R.A. Rankin (1955) The closest packing of spherical caps in n dimensions. Proc. Glasgow Math. Assoc. **2**, 139–144.
139. S. Rippa (1990) Minimal roughness property of the Delaunay triangulation. Comput. Aided Geom. Design **7**, 489–497.
140. S. Rippa (1992) Long thin triangles can be good for linear interpolation. SIAM J. Numer. Anal. **29**, 257–270.
141. S. Rippa and B. Schiff (1990) Minimum energy triangulations for elliptic problems. Comput. Methods Appl. Mech. Engrg. **84**, 257–274.
142. A. Robert (1981) A stable numerical integration scheme for the primitive meteorological equations. Atmosphere-Ocean **19**, 35–46.

143. M. Sabin (2002) Subdivision of box-splines. *Tutorials on Multiresolution in Geometric Modelling*, A. Iske, E. Quak, and M.S. Floater (eds.), Springer-Verlag, Heidelberg, 3–23.
144. M. Sabin (2002) Eigenanalysis and artifacts of subdivision curves and surfaces *Tutorials on Multiresolution in Geometric Modelling*, A. Iske, E. Quak, and M.S. Floater (eds.), Springer-Verlag, Heidelberg, 69–92.
145. A. Said and W.A. Pearlman (1996) A new, fast, and efficient image codec based on set partitioning in hierarchical trees, IEEE Trans. Circuits and Systems for Video Technology **6**, 243–250.
146. R. Schaback (1995) Error estimates and condition numbers for radial basis function interpolation. Advances in Comp. Math. **3**, 251–264.
147. R. Schaback (1995) Creating surfaces from scattered data using radial basis functions. *Mathematical Methods for Curves and Surfaces*, M. Dæhlen, T. Lyche, and L. L. Schumaker (eds.), Vanderbilt University Press, Nashville, 1995, 477–496.
148. R. Schaback (1995) Multivariate interpolation and approximation by translates of a basis function. *Approximation Theory VIII, Vol. 1: Approximation and Interpolation*, C. K. Chui and L. L. Schumaker (eds.), World Scientific, Singapore, 1995, 491–514.
149. R. Schaback (1999) Native Hilbert spaces for radial basis functions I. International Series of Numerical Mathematics **132**, 255–282.
150. R. Schaback (2000) A unified theory of radial basis functions. J. Comput. Appl. Math. **121**, 165–177.
151. R. Schaback (2002) Stability of radial basis function interpolants. *Approximation Theory X: Wavelets, Splines, and Applications*, C.K. Chui, L.L. Schumaker, and J. Stöckler (eds.), Vanderbilt Univ. Press, Nashville, 433–440.
152. R. Schaback and H. Wendland (2002) Inverse and saturation theorems for radial basis function interpolation. Math. Comp. **71**, 669–681.
153. R. Schaback and H. Wendland (2001) Characterization and construction of radial basis functions. *Multivariate Approximation and Applications*, N. Dyn, D. Leviatan, D. Levin, and A. Pinkus (eds.), Cambridge University Press, Cambridge, 1–24.
154. A.E. Scheidegger (1974) *The Physics of Flow in Porous Media*. University of Toronto Press, Toronto.
155. I.J. Schoenberg (1938) Metric spaces and positive definite functions. Trans. Amer. Math. Soc. **44**, 522–536.
156. I.J. Schoenberg (1938) Metric spaces and completely monotone functions. *Annals of Mathematics* **39**, 811–841.
157. L.L. Schumaker (1976) Fitting surfaces to scattered data. *Approximation Theory II*, G.G. Lorentz, C.K. Chui, and L.L. Schumaker (eds.), 203–268.
158. L.L. Schumaker (1987) Numerical aspects of spaces of piecewise polynomials on triangulations. *Algorithms for Approximation*, J. C. Mason and M. G. Fox (eds.), Clarendon Press, Oxford, 373–406.
159. L.L Schumaker (1987) Triangulation methods. Topics in Multivariate Approximation, C. K. Chui, L. L. Schumaker, and F. Utreras (eds.), Academic Press, New York, 219–232.
160. L.L. Schumaker (1993) Triangulations in CAGD. IEEE Comp. Graph. Appl. , January 1993, 47–52.

161. D.S. Scott (1984) The complexity of interpolating given data in three space with a convex function of two variables, J. Approx. Theory **42**, 52–63.
162. J.S. Scroggs and F.H.M. Semazzi (1995) A conservative semi-Lagrangian method for multidimensional fluid dynamics applications. Numer. Meth. Part. Diff. Eq. **11**, 445–452.
163. J.A. Sethian, A.J. Chorin, and P. Concus (1983) Numerical solution of the Buckley-Leverett equations. Paper SPE 12254, Soc. Petrol. Eng. Symp. on Reservoir Simulation, San Francisco, Nov. 1983.
164. D.B. Shmoys (1995) Computing near-optimal solutions to combinatorial optimization problems. DIMACS, Ser. Discrete Math. Theor. Comput. Sci. **20**, 355–397.
165. A. Staniforth and J. Côté (1991) Semi-Lagrangian integration schemes for atmospheric models – a review. Mon. Wea. Rev. **119**, 2206–2223.
166. D. Taubman (2000) High performance scalable image compression with EBCOT. IEEE Transactions on Image Processing, July 2000, 1158–1170.
167. D. Taubman, M.W. Marcellin (2002) *JPEG2000: Image Compression Fundamentals, Standards and Practice*, Kluwer, Boston, 2002.
168. E.F. Toro (1999) *Riemann Solvers and Numerical Methods for Fluid Dynamics*. Springer-Verlag, Berlin.
169. H.J. Welge (1952) A simplified method for computing oil recovery by gas or water drive. *Trans. AIME* **195**, 97–108.
170. H. Wendland (1995) Piecewise polynomial, positive definite and compactly supported radial functions of minimal degree. Advances in Comp. Math. **4**, 389–396.
171. H. Wendland (1998) Numerical solution of variational problems by radial basis functions. *Approximation Theory IX, Volume 2: Computational Aspects*, C. K. Chui and L. L. Schumaker (eds.), Vanderbilt University Press, Nashville, 361–368.
172. D.V. Widder (1971) *An Introduction to Transform Theory*. Academic Press, New York.
173. J.W.J. Williams (1964) Algorithm 232 (HEAPSORT). *Communications of the ACM* **7**, 347–348.
174. I.H. Witten, R.M. Neal, and J.G. Cleary (1987) Arithmetic coding for data compression. Comm. of the ACM **30**, 520–540.
175. Z. Wu (1995) Multivariate compactly supported positive definite radial functions, Advances in Comp. Math. **4**, 283–292.
176. Z. Wu and R. Schaback (1993) Local error estimates for radial basis function interpolation of scattered data. IMA J. Numer. Anal. **13**, 13–27.
177. S. T. Zalesak (1979) Fully multidimensional flux-corrected transport algorithms for fluids. J. Comput. Phys. **31**, 335–362.
178. F. Zeilfelder (2002) Scattered data fitting with bivariate splines. *Tutorials on Multiresolution in Geometric Modelling*, A. Iske, E. Quak, and M.S. Floater (eds.), Springer-Verlag, Heidelberg, 243–286.

Index

Editorial Policy

§1. Volumes in the following three categories will be published in LNCSE:

i) Research monographs
ii) Lecture and seminar notes
iii) Conference proceedings

Those considering a book which might be suitable for the series are strongly advised to contact the publisher or the series editors at an early stage.

§2. Categories i) and ii). These categories will be emphasized by Lecture Notes in Computational Science and Engineering. **Submissions by interdisciplinary teams of authors are encouraged.** The goal is to report new developments – quickly, informally, and in a way that will make them accessible to non-specialists. In the evaluation of submissions timeliness of the work is an important criterion. Texts should be well-rounded, well-written and reasonably self-contained. In most cases the work will contain results of others as well as those of the author(s). In each case the author(s) should provide sufficient motivation, examples, and applications. In this respect, Ph.D. theses will usually be deemed unsuitable for the Lecture Notes series. Proposals for volumes in these categories should be submitted either to one of the series editors or to Springer-Verlag, Heidelberg, and will be refereed. A provisional judgment on the acceptability of a project can be based on partial information about the work: a detailed outline describing the contents of each chapter, the estimated length, a bibliography, and one or two sample chapters – or a first draft. A final decision whether to accept will rest on an evaluation of the completed work which should include

- at least 100 pages of text;
- a table of contents;
- an informative introduction perhaps with some historical remarks which should be accessible to readers unfamiliar with the topic treated;
- a subject index.

§3. Category iii). Conference proceedings will be considered for publication provided that they are both of exceptional interest and devoted to a single topic. One (or more) expert participants will act as the scientific editor(s) of the volume. They select the papers which are suitable for inclusion and have them individually refereed as for a journal. Papers not closely related to the central topic are to be excluded. Organizers should contact Lecture Notes in Computational Science and Engineering at the planning stage.

In exceptional cases some other multi-author-volumes may be considered in this category.

§4. Format. Only works in English are considered. They should be submitted in camera-ready form according to Springer-Verlag's specifications. Electronic material can be included if appropriate. Please contact the publisher. Technical instructions and/or TEX macros are available via http://www.springer.de/math/authors/help-momu.html. The macros can also be sent on request.

General Remarks

Lecture Notes are printed by photo-offset from the master-copy delivered in camera-ready form by the authors. For this purpose Springer-Verlag provides technical instructions for the preparation of manuscripts. See also *Editorial Policy*.

Careful preparation of manuscripts will help keep production time short and ensure a satisfactory appearance of the finished book.

The following terms and conditions hold:

Categories i), ii), and iii):
Authors receive 50 free copies of their book. No royalty is paid. Commitment to publish is made by letter of intent rather than by signing a formal contract. Springer-Verlag secures the copyright for each volume.

For conference proceedings, editors receive a total of 50 free copies of their volume for distribution to the contributing authors.

All categories:
Authors are entitled to purchase further copies of their book and other Springer mathematics books for their personal use, at a discount of 33,3 % directly from Springer-Verlag.

Addresses:

Timothy J. Barth
NASA Ames Research Center
NAS Division
Moffett Field, CA 94035, USA
e-mail: barth@nas.nasa.gov

Michael Griebel
Institut für Angewandte Mathematik
der Universität Bonn
Wegelerstr. 6
53115 Bonn, Germany
e-mail: griebel@iam.uni-bonn.de

David E. Keyes
Department of Applied Physics
and Applied Mathematics
Columbia University
200 S. W. Mudd Building
500 W. 120th Street
New York, NY 10027, USA
e-mail: david.keyes@columbia.edu

Risto M. Nieminen
Laboratory of Physics
Helsinki University of Technology
02150 Espoo, Finland
e-mail: rni@fyslab.hut.fi

Dirk Roose
Department of Computer Science
Katholieke Universiteit Leuven
Celestijnenlaan 200A
3001 Leuven-Heverlee, Belgium
e-mail: dirk.roose@cs.kuleuven.ac.be

Tamar Schlick
Department of Chemistry
Courant Institute of Mathematical
Sciences
New York University
and Howard Hughes Medical Institute
251 Mercer Street
New York, NY 10012, USA
e-mail: schlick@nyu.edu

Springer-Verlag, Mathematics Editorial IV
Tiergartenstrasse 17
69121 Heidelberg, Germany
Tel.: *49 (6221) 487-8185
e-mail: peters@springer.de
http://www.springer.de/math/
peters.html

Lecture Notes in Computational Science and Engineering

Vol. 1 D. Funaro, *Spectral Elements for Transport-Dominated Equations.* 1997. X, 211 pp. Softcover. ISBN 3-540-62649-2

Vol. 2 H. P. Langtangen, *Computational Partial Differential Equations.* Numerical Methods and Diffpack Programming. 1999. XXIII, 682 pp. Hardcover. ISBN 3-540-65274-4

Vol. 3 W. Hackbusch, G. Wittum (eds.), *Multigrid Methods V.* Proceedings of the Fifth European Multigrid Conference held in Stuttgart, Germany, October 1-4, 1996. 1998. VIII, 334 pp. Softcover. ISBN 3-540-63133-X

Vol. 4 P. Deuflhard, J. Hermans, B. Leimkuhler, A. E. Mark, S. Reich, R. D. Skeel (eds.), *Computational Molecular Dynamics: Challenges, Methods, Ideas.* Proceedings of the 2nd International Symposium on Algorithms for Macromolecular Modelling, Berlin, May 21-24, 1997. 1998. XI, 489 pp. Softcover. ISBN 3-540-63242-5

Vol. 5 D. Kröner, M. Ohlberger, C. Rohde (eds.), *An Introduction to Recent Developments in Theory and Numerics for Conservation Laws.* Proceedings of the International School on Theory and Numerics for Conservation Laws, Freiburg / Littenweiler, October 20-24, 1997. 1998. VII, 285 pp. Softcover. ISBN 3-540-65081-4

Vol. 6 S. Turek, *Efficient Solvers for Incompressible Flow Problems.* An Algorithmic and Computational Approach. 1999. XVII, 352 pp, with CD-ROM. Hardcover. ISBN 3-540-65433-X

Vol. 7 R. von Schwerin, *Multi Body System SIMulation.* Numerical Methods, Algorithms, and Software. 1999. XX, 338 pp. Softcover. ISBN 3-540-65662-6

Vol. 8 H.-J. Bungartz, F. Durst, C. Zenger (eds.), *High Performance Scientific and Engineering Computing.* Proceedings of the International FORTWIHR Conference on HPSEC, Munich, March 16-18, 1998. 1999. X, 471 pp. Softcover. 3-540-65730-4

Vol. 9 T. J. Barth, H. Deconinck (eds.), *High-Order Methods for Computational Physics.* 1999. VII, 582 pp. Hardcover. 3-540-65893-9

Vol. 10 H. P. Langtangen, A. M. Bruaset, E. Quak (eds.), *Advances in Software Tools for Scientific Computing.* 2000. X, 357 pp. Softcover. 3-540-66557-9

Vol. 11 B. Cockburn, G. E. Karniadakis, C.-W. Shu (eds.), *Discontinuous Galerkin Methods.* Theory, Computation and Applications. 2000. XI, 470 pp. Hardcover. 3-540-66787-3

Vol. 12 U. van Rienen, *Numerical Methods in Computational Electrodynamics.* Linear Systems in Practical Applications. 2000. XIII, 375 pp. Softcover. 3-540-67629-5

Vol. 13 B. Engquist, L. Johnsson, M. Hammill, F. Short (eds.), *Simulation and Visualization on the Grid.* Parallelldatorcentrum Seventh Annual Conference, Stockholm, December 1999, Proceedings. 2000. XIII, 301 pp. Softcover. 3-540-67264-8

Vol. 14 E. Dick, K. Riemslagh, J. Vierendeels (eds.), *Multigrid Methods VI.* Proceedings of the Sixth European Multigrid Conference Held in Gent, Belgium, September 27-30, 1999. 2000. IX, 293 pp. Softcover. 3-540-67157-9

Vol. 15 A. Frommer, T. Lippert, B. Medeke, K. Schilling (eds.), *Numerical Challenges in Lattice Quantum Chromodynamics.* Joint Interdisciplinary Workshop of John von Neumann Institute for Computing, Jülich and Institute of Applied Computer Science, Wuppertal University, August 1999. 2000. VIII, 184 pp. Softcover. 3-540-67732-1

Vol. 16 J. Lang, *Adaptive Multilevel Solution of Nonlinear Parabolic PDE Systems.* Theory, Algorithm, and Applications. 2001. XII, 157 pp. Softcover. 3-540-67900-6

Vol. 17 B. I. Wohlmuth, *Discretization Methods and Iterative Solvers Based on Domain Decomposition.* 2001. X, 197 pp. Softcover. 3-540-41083-X

Vol. 18 U. van Rienen, M. Günther, D. Hecht (eds.), *Scientific Computing in Electrical Engineering.* Proceedings of the 3rd International Workshop, August 20-23, 2000, Warnemünde, Germany. 2001. XII, 428 pp. Softcover. 3-540-42173-4

Vol. 19 I. Babuška, P. G. Ciarlet, T. Miyoshi (eds.), *Mathematical Modeling and Numerical Simulation in Continuum Mechanics.* Proceedings of the International Symposium on Mathematical Modeling and Numerical Simulation in Continuum Mechanics, September 29 - October 3, 2000, Yamaguchi, Japan. 2002. VIII, 301 pp. Softcover. 3-540-42399-0

Vol. 20 T. J. Barth, T. Chan, R. Haimes (eds.), *Multiscale and Multiresolution Methods.* Theory and Applications. 2002. X, 389 pp. Softcover. 3-540-42420-2

Vol. 21 M. Breuer, F. Durst, C. Zenger (eds.), *High Performance Scientific and Engineering Computing.* Proceedings of the 3rd International FORTWIHR Conference on HPSEC, Erlangen, March 12-14, 2001. 2002. XIII, 408 pp. Softcover. 3-540-42946-8

Vol. 22 K. Urban, *Wavelets in Numerical Simulation.* Problem Adapted Construction and Applications. 2002. XV, 181 pp. Softcover. 3-540-43055-5

Vol. 23 L. F. Pavarino, A. Toselli (eds.), *Recent Developments in Domain Decomposition Methods.* 2002. XII, 243 pp. Softcover. 3-540-43413-5

Vol. 24 T. Schlick, H. H. Gan (eds.), *Computational Methods for Macromolecules: Challenges and Applications.* Proceedings of the 3rd International Workshop on Algorithms for Macromolecular Modeling, New York, October 12-14, 2000. 2002. IX, 504 pp. Softcover. 3-540-43756-8

Vol. 25 T. J. Barth, H. Deconinck (eds.), *Error Estimation and Adaptive Discretization Methods in Computational Fluid Dynamics.* 2003. VII, 344 pp. Hardcover. 3-540-43758-4

Vol. 26 M. Griebel, M. A. Schweitzer (eds.), *Meshfree Methods for Partial Differential Equations.* 2003. IX, 466 pp. Softcover. 3-540-43891-2

Vol. 27 S. Müller, *Adaptive Multiscale Schemes for Conservation Laws.* 2003. XIV, 181 pp. Softcover. 3-540-44325-8

Vol. 28 C. Carstensen, S. Funken, W. Hackbusch, R. H. W. Hoppe, P. Monk (eds.), *Computational Electromagnetics.* Proceedings of the GAMM Workshop on "Computational Electromagnetics", Kiel, Germany, January 26-28, 2001. 2003. X, 209 pp. Softcover. 3-540-44392-4

Vol. 29 M. A. Schweitzer, *A Parallel Multilevel Partition of Unity Method for Elliptic Partial Differential Equations.* 2003. V, 194 pp. Softcover. 3-540-00351-7

Vol. 30 T. Biegler, O. Ghattas, M. Heinkenschloss, B. van Bloemen Waanders (eds.), *Large-Scale PDE-Constrained Optimization.* 2003. VI, 349 pp. Softcover. 3-540-05045-0

Vol. 31 M. Ainsworth, P. Davies, D. Duncan, P. Martin, B. Rynne (eds.) *Topics in Computational Wave Propagation.* Direct and Inverse Problems. 2003. VIII, 399 pp. Softcover. 3-540-00744-X

Vol. 32 H. Emmerich, B. Nestler, M. Schreckenberg (eds.) *Interface and Transport Dynamics.* Computational Modelling. 2003. XV, 432 pp. Hardcover. 3-540-40367-1

Vol. 33 H. P. Langtangen, A. Tveito (eds.) *Advanced Topics in Computational Partial Differential Equations.* Numerical Methods and Diffpack Programming. 2003. XIX, 658 pp. Softcover. 3-540-01438-1

Vol. 34 V. John, *Large Eddy Simulation of Turbulent Incompressible Flows.* Analytical and Numerical Results for a Class of LES Models. 2004. XII, 261 pp. Softcover. 3-540-40643-3

Vol. 35 E. Bänsch, *Challenges in Scientific Computing - CISC 2002.* Proceedings of the Conference *Challenges in Scientific Computing,* Berlin, October 2-5, 2002. 2003. VIII, 287 pp. Hardcover. 3-540-40887-8

Vol. 36 B. N. Khoromskij, G. Wittum, *Numerical Solution of Elliptic Differential Equations by Reduction to the Interface.* 2004. XI, 293 pp. Softcover. 3-540-20406-7

Vol. 37 A. Iske, *Multiresolution Methods in Scattered Data Modelling.* 2004. XII, 182 pp. Softcover. 3-540-20479-2

Texts in Computational Science and Engineering

Vol. 1 H. P. Langtangen, *Computational Partial Differential Equations.* Numerical Methods and Diffpack Programming. 2nd Edition 2003. XXVI, 855 pp. Hardcover. ISBN 3-540-43416-X

Vol. 2 A. Quarteroni, F. Saleri, *Scientific Computing with MATLAB.* 2003. IX, 257 pp. Hardcover. ISBN 3-540-44363-0

For further information on these books please have a look at our mathematics catalogue at the following URL: `http://www.springer.de/math/index.html`

Printing and Binding: Strauss GmbH, Mörlenbach